GRAPHIC CHARTS IN BUSINESS

HOW TO MAKE AND USE THEM

BY
ALLAN C. HASKELL
Author of "How To Make and Use Graphic Charts"

ASSISTED BY
JOSEPH G. BREAZNELL

WITH AN INTRODUCTION BY
RICHARD T. DANA
Consulting Engineer

FIRST EDITION

NEW YORK
1922

PREFACE

This book was written as a companion volume to "How to Make and Use Graphic Charts," the large demand for which the author ascribes mainly to the tremendous public interest in the subject.

The extraordinary growth in the use of graphic methods which has been, perhaps, more voluminous in the last five years than in the whole period prior to 1917, seems to be due to the fact that three of the main functions of charts can be performed by their use much more rapidly, easily and cheaply, than by any other method.

For Computation, whereby many, if not indeed most, of the processes of mathematics can be performed without the labor and often without the knowledge of the mathematics involved; for Recording multitudes of facts of a great variety; and, for the Demonstration of facts in a quantitative way, the graphic chart is the best instrument that has yet been developed. As a consequence, thousands of business concerns have adopted graphics as a standard method for the promotion of greater efficiency in organization, advertising, production control, etc.; colleges and universities are giving special courses in graphic methods; and "graphic service" organizations are springing up for the purpose of making charts for those who need them and cannot learn how to make them fast enough.

This, however, is of no great help to the vast majority of business and professional men who either never went to college or who left it long ago, and who, if not too tired, are at least too busy to plunge into a subject which contains many technical questions, and the current literature of which is "shot full of mathematics."

To make it easy for the man of business to see when and how graphic methods can best serve his purposes and thus to save him money, time or trouble, is the purpose of this book, and in pursuit of this idea mathematical discussion has been avoided. A great many examples, illustrated by actual charts, have been included as more valuable than elaborate discussions in print.

Following the practice in the author's earlier book there is given also a bibliography of the principal books and articles, containing material on business statistics and graphic methods, that have come to notice since the publication of "How to Make and Use Graphic Charts."

It is probably inevitable that with the large growth in the use of charts there is developing an erroneous application of charts to a number of different kinds of business problems. Particularly does this apply in the use of charts with plain arithmetic ruling, which are best adapted for showing a comparison of absolute numerical quantities, in the place of the so-called ratio charts which are best adapted for showing a comparison of percentage variations, comparative trends, etc. As the author has attempted to show in considerable length in this book, many kinds of comparative quantities cannot be correctly shown on any other than

PREFACE

ratio paper, and yet this has been attempted and erroneous results produced in a great many cases, simply because information in regard to these points have not heretofore been broadcasted, as it were, among those interested in charts. To correct this tendency the author has been at considerable pains.

The author wishes to express his sincerest appreciation to the various publishers for the quotations he has made from the periodical press and he has endeavored to give due credit throughout the text, wherever abstracts have been made.

The reader is urged to give careful consideration to Mr. Dana's introductory chapter as he has pointed out there, in an unusually striking manner, why graphic charts perform certain functions better than any other method of statistical presentation. It is because of this fact that their use is growing so rapidly, and it is the author's belief that graphic charts are destined to play as important a part in the records of every well-organized concern as do the well-known and universally used methods of book-keeping.

New York City,
   June, 1922.

CONTENTS

CONTENTS

CONTENTS

CONTENTS

CHAPTER I

The reason for the tremendous increase in the use and application of graphic charts in the last few years is that it is possible to represent in pictorial form, by means of them, the relative magnitudes of quantities and the relative changes that take place with the process of time.

Charts appeal to the mind through the eye which is the organ best adapted by nature for the comparison of quantities, because for countless ages man has been compelled to estimate quantities,—distances, volumes, masses, areas, etc.,—by the eye alone and, much of the time, subconsciously. Our conceptions of space are mostly visual conceptions. We do not grasp the idea of the size of a locomotive, for instance, except by picturing in our mind's eye what a locomotive of a certain size would look like. Our most tangible conception of the value of a sum of money is the size of the pile of gold representing this amount, or the size of the pile of cotton or logs or coal, or any other commodity that we are most familiar with, that such a sum of money would buy.

It is a fact of experience that if we wish to estimate the height of a house, the length of a block, the weight of a pig, or the area of a lot, our estimate will be much closer if we are able to make a direct visual comparison between the unknown object of which it is desired to estimate the dimensions, and another similar object of known dimensions, than by any other method short of direct scientific mensuration involving mathematical and mechanical devices for such comparisons. By this well-known method, experienced drovers readily estimate within a few pounds the weight of steers as they walk or trot past the loading platform, and the same general fact applies to our experience in an enormous number of human activities. Consequently, the best way to convey an impression of quantity on paper to another person is by means of the nearest approximation to an eye view that is practicable, namely—pictures. A well-known illustration of this fact is the use of the picture of a ship to represent the size of a navy, the ship representing the United States navy being two-thirds larger than the one representing the Japenese navy, for example. That is, by making the drawings of the ships indicate such a difference, the relative sizes of the navies, as approved by the Arms Conference, are indicated.

Such pictures are not charts proper but for a long time they offered the best method of illustrating comparative values. Now, naturally, the most accurate method of indicating quantity is by the numerical figures which show its magnitude with respect to an established standard; and by the application of figures to a semi-pictorial representation, a great increase in accuracy was achieved,—the indications of a curved line in conjunction with figures being much more accurate than those of a picture,—and when the short mental step was completed that was necessary to gain a correct conception of the chart as a device for that purpose, great progress immediately began in the art of conveniently illustrating the desired facts and figures in business.

Another advantage derives from the use of charts that is often overlooked by persons who have not made much use of them. This is the fact that the shape of a curve can be remembered very much more easily than the general trend of a column of figures indicating the same thing. The reason for this is that for countless ages man has been compelled to remember what he has seen, to remember what it looked like, to remember its relative shape and contour as well as its relative size, whereas it is only during a comparatively few years, certainly not more than two hundred, since the majority of mankind have had to carry in mind, day after day, thousands of relative quantities expressed by numbers. Psychologists tell us that there are three kinds of memory,—that depending upon the conscious exercise of the visual function, that on the oral function through the ears, and that through the function of movement. People whose memories properly belong in the first class are said to have visual memories, those of the second, oral memories, and people in the third category are said to be motor-minded. Many artists can remember the shapes and colors of the things that they have seen and recreate them later in the studio. Many musicians have extraordinary faculties for remembering combinations of sounds and are not good at remembering colors, shapes, etc. But the great majority of people can appreciate, remember and compare the things they have seen a great deal better than they can those they have heard. Furthermore there is no convenient way of conveying information on paper through the oral function but we have a practically perfect method of doing it through the visual function.

The methods of conveying information through the eye by the shape of curves and pictorially are economically and mentally convenient, and will save our gray matter from wear and tear, since the grasping of the facts, the memorizing thereof, and their subconscious classification, are done without the necessity of many of the mental processes which are indispensible to the employment of figures. The shape of a curve gives us a mental picture of the trend of our income, corresponding to a certain period, without the necessity of our translating figures into the particular mental conception that best fits our own mentality. Secondly, the active man of business can assimilate a good many more figures in the course of a day's work, and remember them, if he does so through the medium of charts than if he does so through tables of figures. As to the above there is no question. It is based upon a well-established psychological law and is being illustrated again and again in all kinds of business activities every day.

It should be appreciated by all of those who are using charts in their work and especially by those who are expanding their use to other functions of their business, as is the case with the great majority of business men today, that no absolute values are secured by eye function alone. The lumbering and clumsy buffalo would look the same size as the sprightly and elusive flea provided that the buffalo were far enough away and the flea near enough at hand. Therefore, in the preparation of a chart, a line of a certain length may indicate anything from more than a million dollars to less than thirty cents, depending upon the scale which must be labeled on the chart. This scale on the chart when comparing figures, takes the place of our perspective sense when applied to the buffalo and the flea.

In this manner what for business purposes we may call absolute values in terms of the dollar, the day, the year, etc., may be, as it were, laid out to their relative sizes by the numbering of the chart, and by remembering our number criterion (and this is a simple matter in most cases) we can remember also the shape of the curve and its location on the paper and thus remember the absolute values that we find necessary with far less labor and fatigue than by the use of figures alone or statements of quantity.

Thirdly, there follows from the above a development, not treated in this book, which is limited to the known mathematical field, of geometrical relations which, applied to the curves in charts, result in the development of numerous processes of computation whereby the mathematical results desired may be obtained without the mental effort or the mathematical training otherwise necessary. Persons interested in this phase of the subject are referred to the book by the same author as this one, entitled "How to Make and Use Graphic Charts."

From all of the above it is very apparent that there is a considerable variety of usefulness obtainable in different kinds of chart forms. The graduations of different kinds make certain impressions upon the eye and are primarily adapted to the work of preparing charts from given figures. Where absolute figures are to be stated for processes of record or memory it is desirable to have equal spaces on the chart to represent equal values in the figures. For business purposes, therefore, the same period of time is always indicated by the same horizontal space on the same chart, that is, a yearly record by weeks would be divided into 52 equal periods of 1 week each. When, however, it is desired to compare the trend of one set of figures with the trend of another set of figures in order to show how one may be affected by the other, the vertical scale of the chart should be that of the so-called "ratio chart" whereon equal percentage variations are indicated by equal spaces upon the paper. Thus, by the use of a chart we can pictorially apply to figures a law of logic known as the law of concomitant variation. This law, considered so important by the logicians and applied many times every day by every business man, is made applicable to figures for purposes of comparison in the most economical way through the use of graphic charts in a manner that has been developed only within the last few years and Mr. Haskell's chapter on the use of the Ratio Chart is especially recommended to the thoughtful attention of all business men.

Probably one of the most important uses of graphic charts, irrespective of the type, is for the development of analytical thinking and investigation. Invariably a chart fairly bristles with interrogations— why is this sudden decrease? or what does that rapid increase portend at this time? etc.—which in many cases require thought, study and careful research to obtain the correct answer.

A few years back it was only the technical man who was trained to use graphic methods and he not very extensively. Today engineers in constantly increasing numbers use this device for computations, designing, estimating, cost analysis, etc., and, in addition, men in nearly every business and profession have found uses for graphic presentation. Industrial executives and managers, advertising men and salesmen, manufacturers, bank officials, lawyers, physicians, farmers, etc., all have learned that graphic methods are vastly superior to any other for a thorough analysis of certain records.

Today managers of all up-to-date plants are presented with daily reports for comparison, and while a good manager may know at a glance whether the record he receives this morning is above or below the average and very likely remembers what it was yesterday and possibly the day before, the best one alive cannot remember the figures for six months past; and it is generally the results over a period which tell the real story. Also, it is difficult for any man to so visualize figures over a back period as to know whether the increases or decreases are more pronounced than have been former ones, whether they extend over a longer time than they should, etc.

The chart which is plotted daily presents as perfect a picture of the events of a business as the correctness and accuracy of the data from which it is made up allow. No chain is stronger than its weakest link and no chart will give facts about a business when compiled from incorrect reports. Provided, however, the material is dependable, the chart presents a graphic picture of the facts as readily analyzed and capable of suggesting constructive ideas as the engineer's or architect's layout of a building, and just as it would be difficult to comprehensively suggest the necessary alterations and revisions if the engineer came in and told what he was doing, or proposed to do, so it is equally perplexing for the manager to understandingly make the essential recommendations when he has to deal with a conglomerate mass of heterogeneous data. Just as the tracing or blue-print collects into a form that is quickly and easily understood all the facts concerning the building, so does the chart collect the facts concerning a business, and both present a record which may be studied and analyzed.

For the purpose of comparing and analyzing statistics, records, etc., it is not necessary that the user of graphic charts be schooled in higher mathematics or the manipulation of drafting materials. He need only understand the method of charting and be able to handle a straightedge and a pencil or pen.

Necessity of Perspective. L. V. Estes, writing in Industrial Management, January, 1920, brings out the value of the graphic chart in obtaining the proper perspective of a business, as follows:

When facts are presented for consideration in graphic form, the whole view of conditions is seen at once. If columns of figures are looked at, the eye sees and the mind grasps only one point at a time. When these same facts are plotted on paper, the eye can see and take in at the one time the whole series. Not only is ease of comprehension gained by this method but the benefit that comes from getting a perspective is also obtained. If we look at the photograph of a twig which is taken from a point very close to that twig and we have no way of comparing its size with something whose size we know, it is sometimes difficult to discover whether it is a photograph of a twig or of a large branch. We are not looking at it in its usual setting or perspective, and hence we cannot place it in its proper relationship to the rest of the things about us. It is just the same in the case of statistics, they must be in their proper relationship if they are to be evaluated at their correct worth. Graphic representation permits us to obtain this proper perspective.

The organization and functioning of a business may be divided into four principal divisions, administration, marketing, production, industrial relations. Each of these divisions is related to each of the others. They cannot be separated one from the other.

One of the unfortunate tendencies of the modern business man is to become so wrapped up in his particular phase of the business that he fails to realize its appropriate position in relation to the other phases. This is not surprising when one considers the complexity of a modern industry, but it is none the less regrettable. It is realized by many progressive business men and many an honest effort is being made to solve the problem.

However, no business should be satisfied to allow its executives to gather all this information in regard to divisions of the business, other than their own, from outside sources. What the executives need is accurate and easily understandable information about what is going on in the several divisions of their own company. Typewritten reports and columns of statistics about the sections of a company in which the executive in question is not personally interested are almost always too dry and uninteresting to cause the head of the section in question to take the time and the trouble to dig out the information that he ought to have. But when this same information is presented in the form of a simple chart or graph which does not need a lot of time for its comprehension, the busy executive will take the time necessary to study something that is at one and the same time so interesting to him and so helpful.

CHAPTER II

WHAT IS A GRAPHIC CHART?

Figures are the quantitative expression of facts. Graphic charts are the pictorial representation of figures.

Generally speaking, graphic charts furnish a method of representing quantities by means of straight lines or areas, the length of the lines or the amount of the areas being a measure of the value of the quantities. Graphic methods are unusually valuable in the interpretation of tabular statistics, not only because charts are easily and cheaply made, but also because they furnish a quick and comprehensive means of perceiving the variations and trends in the tabular statistics.

Most of us, in examining long columns of figures for purposes of comparison, are able to remember only a limited number of outstanding facts, and beyond a certain point we are very apt to become confused or, at least, to forget some important points in an endeavor to retain new details as they become evident.

By putting our tabular statistics on a graphic chart we get a complete mental picture of all the facts expressed in the figures. Without mental effort and without the risk of forgetting important details, we are able to analyze the figures as represented on the chart. Unusual features are made apparent at once; and by the heights and depths of the "peaks" and "valleys" we are able to determine the relative importance of variations in magnitude. Many an unsuspected condition has been brought to light by making a graphic chart from a series of reports or data on file in an office, and many a loss has been averted as the result.

To the business man the graphic chart presents a method for obtaining a pictorial representation of what transpires in the operation of his business.

It enables him to see at a glance just how his income, expenses, sales, units of output, production costs, labor turnover, and all the other things which he must know about, are varying from day to day, week to week, month to month, and year to year.

When an event depends upon or affects another, or several others, a graphic chart shows pictorially exactly to what extent the one affects the other, or others.

Because the graphic chart presents a complete picture of the course of events—a picture which makes unusual conditions evident at once —this method furnishes a means for executive control which is unequaled by any other.

The graphic chart, by reason of the fact that over a given period of time it shows the trend or tendency of events, may be used as a means of predicting what is apt to occur in the future. This, of course,

6

assumes that the conditions governing future events are the same as those which governed past events. A graphic chart, per se, is not able to look into the future and tell of unforseen occurrences.

The use of graphic charts may be upon as small or as large a scale as is desired. A system may be inaugurated to cover a few important general facts and kept up to date with the aid of a bright stenographer or clerk; or, it may be expanded to include every branch and department of an industry, perhaps developing into departmental size.

The Functions Graphic Charts Perform. There are a great many functions which graphic charts may be utilized to perform. Many of these uses, such as computation, the solution of formulæ, etc., have been employed for some time by the more or less technically trained. So far as the business man is concerned, however, he is much more apt to be interested in their application to the comparison and analysis of his data on production, sales and advertising or of his financial statements of income, expenses and profit. With respect to these things the principal functions of graphic charts are:

1. To show comparisons of magnitude; i. e., to give a picture of actual numerical variations, such as whether our income is $1,000 more this month than it was last month, or whether we produced 75 units this month as compared with 100 last month, etc.

For this purpose the bar chart and the line chart, drawn on plain ruling, are best adapted.

2. To show comparisons of relative increase or decrease; i. e., to give a picture of percentage variations, such as whether our income is 10 per cent. more this month than it was last month, or whether we produced 25 per cent. less this month than last, etc.

For this purpose the line chart drawn on ratio ruling is best adapted.

3. To show comparisons of component parts; i. e., where the sum of several parts equals a whole, to give a picture showing the relation of each part to the other parts and to the whole.

For this purpose subdivided bars, the circular percentage chart, and the trilinear chart are best adapted.

CHAPTER III

KINDS OF GRAPHIC CHARTS FOR BUSINESS PURPOSES

There are a great many types of graphic charts and they may be used for many different purposes. It is the aim of this book, however, to show how to make and use graphic charts pertaining directly to everyday business problems.

Graphic charts for business purposes may be divided into the following classes.

1. Line Charts.
2. Bar Charts.
3. Circular Percentage Charts.
4. Organization Charts.
5. Trilinear Charts.
6. Probability Charts.

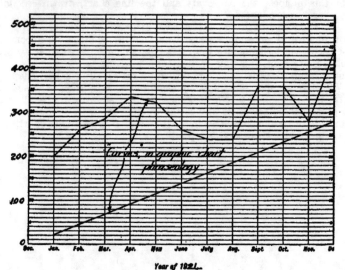

Fig. 1—A Line Chart

Line Charts. With this type of chart the data are shown by a line. The line may be one continued straight line, or it may progress with irregularity, now rising, now falling. This line, whether straight or irregular, is called a "curve." See Fig. 1.

The next chapter explains in detail how a line chart is made.

8

Bar Charts. As suggested by the name, on this type of chart the data are graphically represented by a series of bars, either vertical or

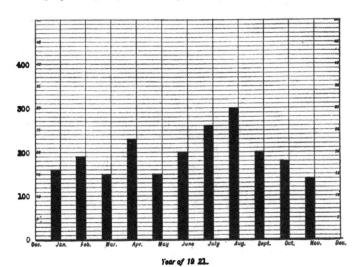

Year of 19 21.

Fig. 2—A Bar Chart

horizontal. Fig. 2. illustrates a vertical bar chart. See Chapter XIII.

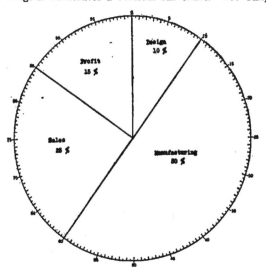

Fig. 3—A Circular Percentage Chart

Circular Percentage Charts. In this type of chart a complete circle represents a unit or a whole. The sectors, into which the circle may be divided, represent the parts into which the whole may be divided. For convenience, the circumference of the circle is divided into 100 equal parts. Then the circle represents 100% and the sectors, drawn to scale, graphically represent the percentage which each bears to the whole. Fig. 3 illustrates the circular percentage chart. Chapter XIV explains it in greater detail.

Organization Charts. The organization chart is a chart, or diagram,

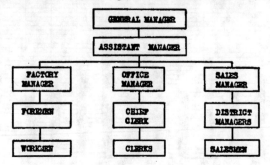

Fig. 4—An Organization Chart

by which the relation of one officer, function, or department, of a company to another is graphically represented. Fig. 4 shows a simple organization chart. See Chapter XV for a more complete discussion.

Trilinear Charts. The trilinear chart is in the form of a triangle. It is illustrated by Fig. 5.

The use of the trilinear chart in business is explained in Chapter XVI.

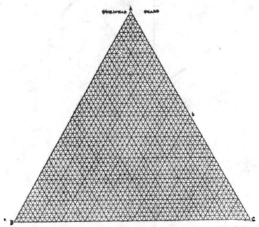

Fig. 5—A Trilinear Chart

Probability Charts. A frequency curve, when plotted upon ordinary cross-section paper, has the so-called bell-shape. Probability Paper is so designed that if a frequency curve is plotted upon it, the curve will

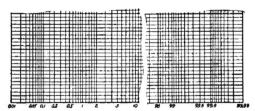

Fig. 6—Probability Paper

be in the form of a straight line. This is explained at **greater length** in **the** chapter on Probability Charts. Fig. 6 shows a **section of** Probability Paper. The full-sized sheet is 8½ by 11 in.

CHAPTER IV

How a Line Chart is Made

Business records of all kinds are usually kept with reference to some period of time—by days, weeks, months, or years. For purposes of comparison the time element also usually enters. We want to know, for example, how the gross income from our business compares this year with last year, or we desire to know whether our production costs are relatively greater or less with respect to our output this month than they were last month, and so on.

Let us assume that we operate a small business, the gross income from which averages about $400 per day. We have been in the habit, we will say, of keeping a record of our daily receipts and we would like to make from this record a graphic chart of last week's business.

Our receipts were as follows:

Monday	$350.
Tuesday	300.
Wednesday	400.
Thursday	450.
Friday	500.
Saturday	450.

The first step is to show, graphically, a week divided up into days; that is, to make a picture which shall represent the daily lapse of time over a period of one week.

This is done by drawing a horizontal line of any convenient length to represent the period of one week. Then, along this line, at equal distances apart, make dividing lines to represent the days of the week, Monday to Saturday. The dividing lines are made equal distances apart to symbolize equal intervals of time (the days). Next, mark the dividing lines with the names of the days of the week, Monday to Saturday, and we have Fig. 1 which is a pictorial representation of a period of one week divided into days.

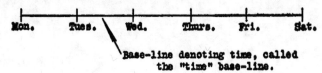

Base-line denoting time, called the "time" base-line.

Fig. 1—One Week Divided into Days

Fig. 2 shows what the receipts were for each day, as indicated in the table, but this has no particular advantage over the table in en-

abling us to visualize the relative magnitudes of the amounts taken in
on the various days.

But now look at Fig. 3. One glance shows upon which day the
receipts were the largest, when they were smallest and about how they
averaged through the week. This is accomplished by *showing the re-*
ceipts graphically.

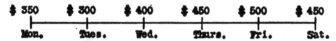

$ 350	$ 300	$ 400	$ 450	$ 500	$ 450
Mon.	Tues.	Wed.	Thurs.	Fri.	Sat.

Fig. 2—The Receipts Hooked-up with the Days

A packet of bills representing an amount equal to $50, is taken
as the unit of measurement. Above the dividing line for each day
is piled the number of packets or units required to make up the receipts for that day. Monday's receipts were $350, therefore
seven unit packets of $50 each are shown to represent the total.
Tuesday's receipts were $300 and six unit packets are shown, and
so on.

It would be a tedious performance, however, if we were always obliged to draw such elaborate pictures to express our facts. This necessity is removed by the simple expedient of drawing a vertical line
at the left extremity of the "time" base-line, and dividing it into
equal parts. We will let each part or division represent $50. This
vertical line with its divisions is called the "scale," and it is with

Fig. 3—The Receipts Shown Graphically

reference to this scale that the receipts for each day are indicated
on the chart, Fig. 4. That is, starting with zero, one division up on
the scale would correspond to one packet of bills and would equal $50,
two divisions would equal $100. So on up the scale—0, $50, $100,
$150, $200, $250, etc., until the scale range is sufficient to include the
largest number to be charted, which, in this case is $500 on Friday.
We will mark our scale "Receipts in Dollars," so that there may be no
doubt as to what the divisions and figures represent, Fig. 4.

Now that we have a scale of Receipts in Dollars to refer to, it is
not necessary to show the packets of money. Instead, directly above

the dividing line for each day, as indicated on the "time" base-line, we will make a dot or point at the proper height with reference to the scale of Receipts in Dollars. That is, above Monday we will make a point at a height, such that if it is referred to the scale of Receipts in

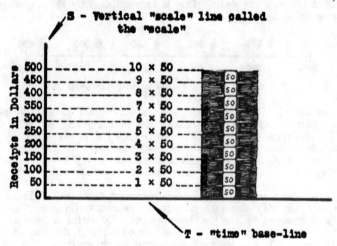

Fig. 4—How the Scale is Arranged

Dollars it will compare with $350 on that scale. Above Tuesday we place a point at a height of $300, above Wednesday at a height of $400, and so on. See Fig. 5.

The function of a well designed graphic chart sheet for business statistics is to furnish a sheet so laid out with cross-sectional lines (vertical and horizontal lines crossing one another) that the labor of locat-

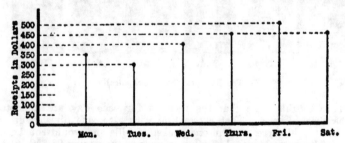

Fig. 5—Daily Amounts Located with Reference to the Scale

ing the dots or points upon it, in their proper positions, is reduced to a minimum. Therefore, for our purpose of locating a record of daily variations for one week, a sheet ruled as in Fig. 6 would be well adapted. It is a very simple matter to place our dots, or plot our points, as it is called, on this sheet and it is equally simple, in reading the chart, after the points have been plotted, to obtain the value of any one point from the vertical scale by means of using the horizontal lines as guides.

The points as plotted on Fig. 6 give us an excellent picture of the relative amounts of our daily receipts, but by connecting these points with straight lines, we obtain a picture which is more complete and

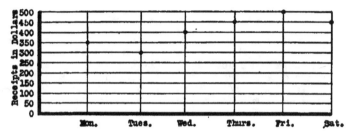

Fig. 6—Graphic Chart Ruled for Weekly Record

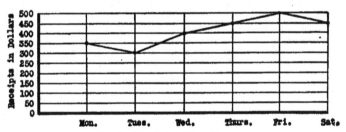

Fig. 7—Completed Line Chart

which furnishes a means of more quickly grasping the situation as a whole. The whole line made up of the short lines connecting the points is called the "curve" of Receipts. Fig. 7 shows the completed graphic chart. It is a simple one, but the principle described may be applied to the making of line charts for all business statistics.

CHAPTER V

THE PLAIN AND RATIO RULINGS

We now come to what we consider the most important thing in connection with the line chart for business statistics. That is the selection of the ruling best adapted to show the facts that you wish to show.

By far the greater part of the line charts that one sees in books, magazines, etc., and even in the records of business concerns, present an incorrect and misleading picture of the facts which they are intended to present. This is because they have not been plotted upon a sheet with ratio ruling.

Let us emphasize this point. We cannot urge too strongly that the ratio ruling—what it is and what it will do—be thoroughly understood. For only then will its advantages and the many "stunts" that can be done with it be apparent.

Do not be turned aside by the thought, gained, perhaps, by a superficial glance at it, that the ratio chart is too complicated and technical, and that it is much simpler and easier to use plain ruling. In the first place there is nothing difficult to understand about the ratio ruling. A knowledge of mathematics is not necessary and any one who can make a chart on plain ruling can make one on ratio ruling. In the second place, it is not a question of whether the plain ruling seems simpler or whether one is more familiar with it. It is a question of whether the charts are correct or incorrect, whether they give a true exposition of the facts which they are intended to give and whether the conclusions drawn from the charts have been based upon false premises or true ones.

We do not wish to detract in any sense from the value of the well-known and simple plain ruling. We only desire to urge an equal understanding of the less known but equally simple ratio ruling.

Plain Ruling. When the lines which run horizontally across the sheet from the vertical scale are spaced equal distances apart, the ruling is called "Plain Ruling." Other names are "rectilinear ruling," "arithmetic ruling," and "difference ruling." A of Fig. 1 shows the lines spaced equidistantly from 0 to 20.

Ratio Ruling. When the lines which run horizontally across the sheet from the vertical scale are spaced logarithmically, the ruling is called "semi-logarithmic," or "ratio ruling." See B of Fig. 1.

To use the ratio ruling it is not necessary to understand the theory of the logarithmic or ratio ruling.[1] All one needs to know are its few peculiarities, described as follows.

Plain and Ratio Rulings Compared. The main difference between the plain and ratio rulings is as follows: that on the rectilinear chart

[1] For those who are interested in the theory, Haskell's "How to Make and Use Graphic Charts" gives a brief and simple explanation.

16

with plain ruling the same *numerical* difference is always represented by the same vertical distance, whereas on the ratio ruling the same *percentage* difference is always represented by the same vertical distance. In other words the plain ruling is adapted to making a comparison of differences in magnitude and the ratio ruling is adapted to making a comparison of differences in percentage. Hence the plain chart is sometimes called the "arithmetic chart," and the other is called the "ratio chart."

A glance at Fig. 1 will help to make the distinction clear. On A, the plain ruling, successive numbers, 0 to 20, occur at equal intervals

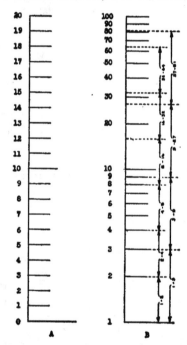

Fig. 1—Comparison of Plain and Ratio Scales

and the numerical difference between equal intervals is always the same. That is, the interval or distance between 1 and 2 is the same as that between 19 and 20, and the numerical difference, 1, is the same in both cases. The interval between 1 and 3 is the same as that between 3 and 5, and the numerical difference, 2, is the same in both instances.

On B, the ratio ruling, the intervals between successive numbers are not equal, but grow progressively less as the numbers get larger. Upon measuring the scale B it will be found that the intervals or distances are the same between numbers which bear a given ratio to one another. That is, the interval between 1 and 2 is the same as that between 2 and 4, 4 and 8, 8 and 16, 16 and 32, etc., and the ratio of

one number to the other is the same in each case. The interval between
1 and 3 is the same as that between 3 and 9, 9 and 27, 27 and 81, etc.,
and the ratio or percentage of one number to the other is the same in
each case.

In order to vividly bring out the difference between the pictures
gained by the use of the two rulings, we have inserted Fig. 2 and 3.
Let us assume that in 1910 we had $1.00 in a certain fund. The first
year following, 1911, the amount was quadrupled, $4.00; the second
year it was divided by two, $2.00; the third year this amount ($2.00)
was quadrupled, $8.00; the fourth year this amount ($8.00) was divided
by two, $4.00; and so on. Tabulated, the data would be as follows:

Year	Amount				
			Amount at Beginning		
1910	$ 1.00				
1911	4.00	400%	increase	over	1910.
1912	2.00	50%	decrease	"	1911.
1913	8.00	400%	increase	"	1912.
1914	4.00	50%	decrease	"	1913.
1915	16.00	400%	increase	"	1914.
1916	8.00	50%	decrease	"	1915.
1917	32.00	400%	increase	"	1916.
1918	16.00	50%	decrease	"	1917.
1919	64.00	400%	increase	"	1918.
1920	32.00	50%	decrease	"	1919.

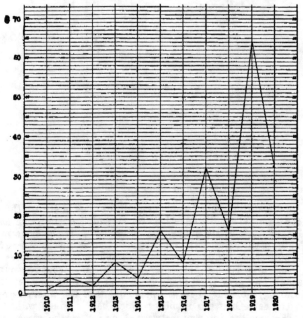

Fig. 2—Graphic Chart of an Amount Alternately Increasing and Decreasing
by the Same Percentage. Plain Ruling.

Thus the amounts alternately increase and decrease by the same percentages. They are quadrupled or increased by 400 per cent., on the odd years and divided by two or decreased by 50 per cent., on the even years. Fig. 2 shows these data plotted on the plain ruling.

The impression gained from this chart is that the amounts are increasing and decreasing at an increasing rate, because the rises and falls in the curve are greater at each step, as the years progress. This, of course, is not the case, for the rates of increase and decrease are the same. The percentages of rise and fall are the same for all the oscillations.

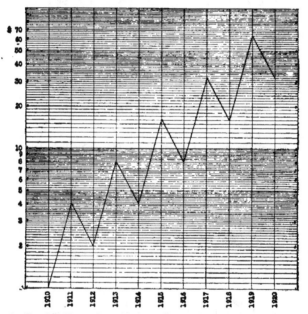

Fig. 3—Graphic Chart of an Amount Alternately Increasing and Decreasing by the Same Percentage. Ratio Ruling.

Now look at Fig. 3. Here we find that the oscillations are all uniform. There is exactly the same rise on the odd years and exactly the same fall on the even years, which is in accordance with the facts.

In this figure the data are plotted on the ratio ruling.

This example should make it perfectly plain that it is absolutely necessary to have an understanding of the functions of both the plain and ratio rulings if one is to obtain a true picture of the facts he desires to know. The use of the plain ruling, when a picture of relative or percentage variations was desired, would result in an incorrect representation of the facts and these, in turn, might lead to false conclusions.

CHAPTER VI

WHEN TO USE THE PLAIN RULING

When you wish merely to show the numerical variations in one set of data the plain ruling is quite satisfactory. Or, if you wish to show the numerical variations in several sets of data and if the limits of variation in each set are the same, or very nearly the same, and you are concerned merely with the numerical amount of variation and not with the percentage of variation, the plain ruling will answer.

H. Caldwell, in an article, "Ratio Chart vs. Difference Chart," in the May, 1920, issue of Business Methods, gives an outline of the uses of the plain ruling as follows:

"The best that can be said for the ordinary cross section paper (graphic charting sheet—plain ruling) or 'Difference' method of charting, is that it always shows the **Difference** in comparisons of magnitudes or values and shows whether there has been an increase or decrease. The base or zero line gives a method of plotting positive and negative quantities, also this method allows a simple comparison of vertical elevations above a base line. The zero line should always be shown in using this method, or else its omission may cause the fluctuations to appear to be greater than is actually the case. In the difference chart, however, comparisons between curves are often meaningless, especially when the curves represent largely unequal magnitudes. When the quantities compared are of nearly the same magnitude the comparative fluctuations are satisfactorily shown on this chart."

Therefore, use the plain ruling when you want to compare one numerical value with another. For example, use the plain ruling when you make a graphic chart of what takes place in your stockroom. There you are dealing with quantities only. You have 10,000 units on hand January 1. You sell 4,000 during January. You sell 6,000 during February and you add 12,000 to stock, etc. Your chart shows you the actual number you have on hand at any time.

Also use the plain ruling for a progress chart. For example, you decide that you will turn out 5,000 units in January, an average of 200 per working day. Then chart on plain ruling your actual performance from day to day in order to know whether you are up to schedule.

CHAPTER VII

WHEN TO USE THE RATIO RULING

When you wish to show percentage variations use the ratio ruling. The point is this: do you want to show the numerical variations, i. e., the difference in dollars, in your revenue and expenses, or do you want to know whether the percentage of increase or decrease in one is greater or less than in the other? Most charts are made with the avowed intention of showing percentage variations, but when drawn on plain ruling they give an absolutely incorrect and misleading picture.

Table I

Year	Advertising Expense	Expense of Salesmen	Salary of Salesmen	Total Sales Cost	Total Sales
1910	$13,500	$35,000	$65,000	$113,500	$525,000
1911	14,000	19,000	67,000	100,000	590,000
1912	18,500	22,000	69,000	109,500	700,000
1913	12,000	17,000	55,000	84,000	730,000
1914	11,500	36,000	58,000	105,500	630,000
1915	17,000	28,000	62,000	107,000	750,000
1916	16,000	27,000	73,000	116,000	825,000
1917	14,000	40,000	68,000	122,000	950,000
1918	14,500	32,000	67,000	113,500	800,000
1919	15,500	36,000	70,000	121,500	1,000,000

Where you are interested only in the numerical variations in one or more sets of data, the Plain chart is the better one to use.

If, however, you wish to know the percentages of variation in one or' more sets of data, and the relative effect of the changes in certain sets of data upon the changes in other sets of data, always use the Ratio chart. The numerical values also may be read on the ratio chart, by one familiar with it, just as easily as on the Plain chart.

The data in Table I have been charted on Fig. 1, Plain ruling, and on Fig. 2, Ratio ruling. Note the following facts: In Fig. 1 it appears as though there were extreme fluctuations in Total Sales and very little variation in the other items. In Fig. 2 the fluctuation of Total Sales is much less marked and that of the other items much more so. This is a correct picture of percentage variations and it will be noted that actually there has been much more variation in Expenses of Salesmen, Advertising Expense, etc., than in Total Sales. For example, the Total Sales were $825,000 in 1916 and $950,000 in 1917, an increase of 15.1 per cent. For the same years the Expenses of the Salesmen were $27,000 and $40,000, an increase of 48.1 per cent. Yet in Fig. 1 the increase of Total Sales is represented by a vertical rise of 12.5 spaces and that of Expenses of Salesmen by 1.3 spaces. Thus, while the percentage of increase of Total Sales was much less than that of Expenses of Salesmen, it shows in Fig. 1 as though it were nearly 10 times as great. In Fig. 2 you will note that there is a much sharper rise, between 1916 and 1917 in the Expenses of Salesmen curve than in the Total Sales curve,—a correct picture of the facts.

Another advantage of the ratio chart is its scale range. This depends upon the number of cycles of logarithmic ruling. Two cycles give a scale range of 1 to 100; 10 to 1,000; 100 to 10,000, etc. (10,000 to 1.000,000 in Fig. 2.) Three cycles give a range of 1 to 1,000; 10 to 10,000, etc. Five cycles give a range of 1 to 100,000; 10 to 1,000,000, etc.

You will note also that while curves b, c, d and e are pretty well jumbled together in Fig. 1, they are spread out in Fig. 2, and whereas in Fig. 1 it is difficult to read the values of the various points with any degree of accuracy, it is easy in Fig. 2. Even in the upper ranges, as in curve a-a, the values may be read nearly as accurately in Fig.

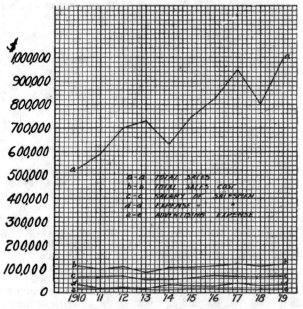

Fig. 1.—Data in Table I Charted on Plain Ruling

2 as in Fig. 1. Thus, yet another advantage is to be noted for the Ratio ruling. It is far superior to the Plain ruling for making comparisons when very small and very large quantities must be considered at the same time. This applies also to cumulative curves, historical curves, etc., where the beginning is small, and from this the amounts increase to large proportions at the end. On the Ratio chart both beginning and end are easily plotted and easily read.

We consider it essential that the reader should have a clear understanding of the use of the Ratio ruling. Therefore, even though certain facts are repeated therein, we are going to include abstracts from two articles in the belief that, because they explain the Ratio chart in different language, some points not clear to the reader in the preceding explanation may be made so in one or the other of the following.

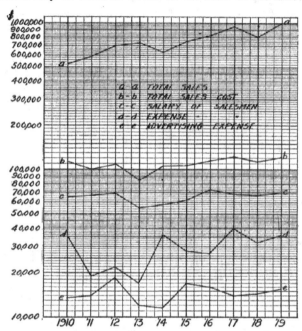

Fig. 2—Data in Table I Charted on Ratio Ruling

To quote further from Mr. Caldwell's article, Business Methods, May, 1920: When the quantities to be charted differ greatly in either value or volume, the wide difference in the amount of vertical movement of the curve, if plotted on the plain ruling, is very misleading and a

Table II

	1902	1903	1904	1905	1906	1907	1908	1909	1910	1911	1912	1913	1914	1915	1916	1917	1918
Estab. Charges	14	17	15	11	10	14	13	15	14	16	14	12	14	12	8	6	3
Mtce. of Plant	11	8	8	9	11	6	7	8	8	5	8	9	6	6	5	4	5
Mtce. of Bldgs.	3	3	1	1	3	2	1	2	2	2	3	3	1	2	2	2	2
General Expenses	26	25	25	24	27	26	27	26	24	25	24	25	26	36	28	30	32
Distribution	46	47	51	55	49	52	52	49	52	52	51	51	53	54	57	58	58
TOTAL %	100	100	100	100	100	100	100	100	100	100	100	100	100	100	100	100	100

source of confusion. It is then that the graphic chart sheet with the ratio ruling is by far the better method of representation of such data.

The ratio chart compares magnitudes or values by their **ratios** and not by their **differences**. This chart shows fluctuations in small magnitudes as clearly as in great ones and, moreover, for comparing in detail any two curves on ratio charts, we may do what we cannot do on difference charts—move bodily either curve until the two are close together. In Fig. 3 the curve, Distribution, appears to fluctuate more violently than any other. This is simply because it represents a larger percentage value. The curve, Maintenance of Buildings, seems to fluctuate the least, and it does so conversely, because it represents the smallest percentage of value. In Fig. 4 the data of Table II are plotted on semilogarithmic or ratio paper, and show the various fluctuations in their true proportions. It is at once seen that the most violent of these fluctuations occur in the Maintenance of Buildings curve, and that the least occur in the Distribution curve. These facts were not brought out in this manner in the difference chart.

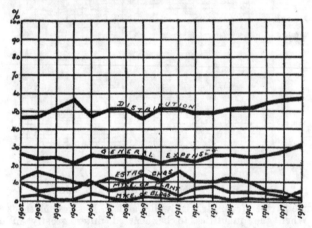

Fig. 3.—Data in Table II Plotted on Plain Ruling

To find the ratio of increase or decrease between any two points on the same or different curves on the ratio chart, we simply measure the vertical distance between them with a pair of dividers, and compare with the vertical scale above the 1. If the distance measured is equal to the distance between the 1 and 2 (or 10 and 20 or 100 and 200), the upper point is just twice the magnitude of the lower, or a 100 per cent. difference. Each line between the 1 and 2 represents an increase or decrease of 10 per cent. The distance between the 2 and 4 is the same as between the 1 and 2, but there being twice as many lines, each line represents 5 per cent. increase or decrease. Curves having an increase from 1 to 2, 2 to 4, 3 to 6, or any other distance between any number on the scale and double that number have an increase of 100 per cent. For instance, in Fig. 4 if we measure on the Maintenance of Buildings curve the vertical distance between points 1902 and 1907 and compare with the vertical scale above the 1, we find that our dividers reach the fifth line above this point, thus representing an increase of 50 per cent.

Equal vertical distances on any number of curves represent equal ratios of increase or decrease. All executives will realize that it is often just as important to watch and control those curves representing materials having a relatively small consumption as it is to watch and control those curves which represent materials of which large quantities are used.

All executives and others interested in graph work should procure a few sheets of these papers and chart some of their data, especially if they have been using the difference methods of charting. The results will surprise them, as they did one business man of whom I heard. His

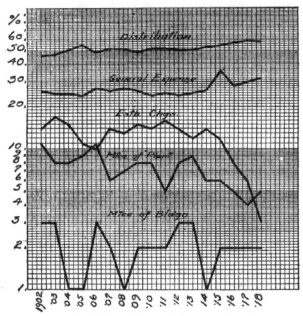

Fig. 4.—Data in Table II Plotted on Ratio Ruling

attention had been recently called to the advantages of the ratio chart, and he recharted his business statistics. He was startled to discover how he had been misled by the difference method.

The Ratio Chart and its Application. Percy A. Bivins, in Industrial Management, July, August, September and October, 1921, wrote an excellent series of articles on the ratio chart, from which the following has been abstracted.

If two quantities, as say 1 and 4, are each doubled at regular intervals and the results plotted in conventional form, their "curves" appear as in Fig. 5. A natural inference from the inclination of the curves, and without reference to the figures on the diagram, would be that—

a—Quantity B has a more rapid rate of increase than Quantity A because its curve is steeper.

b—The rate of increase of each quantity is more rapid towards the end of the period than at the start. This inference follows from the fact that the curves are increasing in steepness.

These inferences are readily seen to be incorrect from the statement of the problem, that in each case the rate is uniformly one of doubling; there is, therefore, no change in rate. It must be borne in mind that the discussion here is not of *magnitudes* but of *rates*. The increase from 1 to 2 is 1; from 2 to 4, it is 2; from 4 to 8, it is 4; and from 8 to 16, it is 8, representing changes in *magnitude* by increase of 1, 2, 4 and 8 respectively but the ratio is uniformly as one to two or in other words, the increase is uniformly at the rate of 100 per cent.

Now if the same amounts are plotted on rulings spaced as shown in Fig. 6 the curves will take the form of straight lines and also will be parallel. If the principle is stated, and accepted, that with such spacings

a—A straight line indicates a uniform rate or ratio, and

b—Parallel lines indicate equal rates or ratios

then it is apparent that curves A and B indicate two amounts which are each uniformly increasing at the same rate.

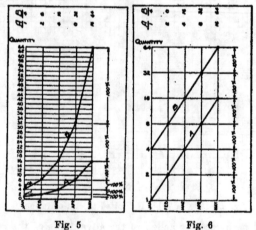

Fig. 5   Fig. 6

To arrive at the reasons for this, it will be noted that in Fig. 5, the horizontal lines representing the magnitudes 0, 2, 4, 6, 8, 10, 12, etc. up to 64 are spaced equidistantly but that the vertical interval or rise between 2 and 4 is one-half of that between 4 and 8 and one-fourth of that between 8 and 16. In Fig. 6, the vertical intervals representing the rises from 1 to 2, 2 to 4, 4 to 8, 8 to 16, 16 to 32 and 32 to 64 are equal. Each of these intervals, therefore, indicates an amount doubled on itself or amounts increasing at a ratio of 1 to 2, or at a rate of 100 per cent. The ratio or percentage spacings being equidistant throughout, points plotted on them at regular distances apart laterally must consequently be straight lines.

In this article the former ruling, illustrated by Fig. 5, is called "arithmetic" ruling, and the latter ruling, illustrated by Fig. 6, is called "ratio" ruling.

Applying these two types of rulings to a typical industrial problem, in Fig. 7 are shown the comparative volumes of sales of several products, charted on arithmetic rulings. The deductions from these graphs are that Product A is fluctuating widely in its volume of sales; that Product B has a moderate fluctuation; and that Product C, where the variations are too small to be read accurately, is fairly uniform.

The same data plotted on the ratio scale as in Fig. 8 show the curves of the three products to be exactly alike. The products are therefore in exactly the same ratio from month to month. This fact may be checked from the tabulation.

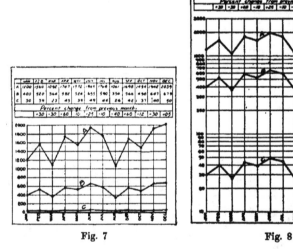

Fig. 7   Fig. 8

The principles of the ratio ruling can best be demonstrated by comparative examples. In Fig. 9, on arithmetic ruling are shown curves of two funds—A and B.

The interpretation at first glance that Fund B is advancing at a more rapid rate than Fund A is not justifiable. The fact is that both curves represent the same rate. Fund A is one of $10.00 and Fund B of $100.00—each compounded at 20 per cent. annually.

When plotted on the ratio scale as in Fig. 10, the graphs of both funds are straight and parallel. The visual impression and the correct one is, therefore, that they represent equal rates.

These differences in the shapes of the curves are brought about by the spacings of the rulings. On the arithmetic scale, in Fig. 9, the vertical distance between ruling 100 and ruling 200 is one-half of that between 200 and 400, and one-fourth of that between 400 and 800.

Each of these sets of figures is in relation of 1 to 2. The spacing of the last two, in reference to the first, is, however, in relation of 1 to 2 and 1 to 4.

On the ratio ruling (Fig. 10) the vertical distance between 100 and 200 is the same as between 200 and 400 and between 400 and 800. There is also an equal distance between 10 and 20, 20 and 40, 40 and 80 or any other combination of figures having a relation of 1 to 2. Similarly if the relations are 1 to 3, as 10, 30, 90 or 100, 300, 900 the spacing between the rulings is equal.

In like manner percentage changes of any amount have the same spacing, no matter whether based on large or small figures. Thus a 25 per cent. increase over 80 is 100 and is shown by the lines 80 and 100. A 25 per cent. increase over 400 is 500, or of 4,000,000 is 5,000,000. On the ratio rulings, their spacings are equal. On the arithmetic scale, however, the 400-500 interval is 5 times as great as between 80 and 100.

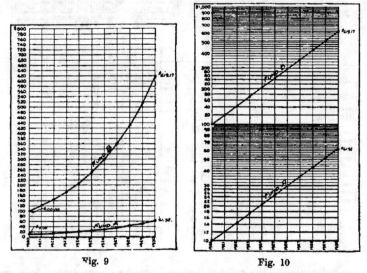

Fig. 9   Fig. 10

Were the amounts 4,000 and 5,000, the interval would be 50 times as great.

In other words, on the ratio scale, the spacing of the lines representing the figures or magnitudes is the same for every fixed relation of the figures. On the arithmetic scale, the spacings increase as the figures grow larger. Curves or "graphs" of the same *rate* of increase and for equal periods or durations of time, when plotted on the ratio scale, are parallel—irrespective of the size of the figures. Graphs representing the same rate and for equal periods when plotted on the arithmetic scale increase in inclination with the increase of the magnitudes. Decreases in rate are similarly affected. Rates or percentage variations are therefore comparable on "ratio" rulings but not on "arithmetic" rulings.

The amounts A and B given in the tabulation in Fig. 11 are plotted as in the chart. The computed percentages of changes from the previous month are noted for convenience on the slopes of the monthly graphs and also on the vertical interval between the points representing the amounts. The steeper inclinations indicate the greater increases or decreases. Equal percentages of change are indicated by lines of equal inclination. The vertical intervals between the points representing the amounts which have equal changes in percentage are also equal. This can be demonstrated by measuring them.

The fact that increases and decreases of the same rate are not represented by the same angles of slope or by equal vertical intervals is due to the "base." Thus an increase from 100 to 200 is an increase of 100 per cent. A decrease from 200 to 100 is a decrease of 50 per cent. The base is 100 in one case and 200 in the other.

The same data represented for comparison on the arithmetic scale are shown in Fig. 12. It is obvious that no perception of rates can be gained from the slopes or rise of the curves except by calculation.

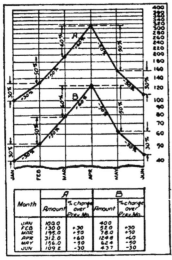

Fig. 11

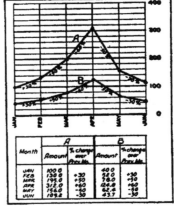

Fig. 12

Ratio charts have certain advantages over charts on the arithmetic basis but also disadvantages. In some cases they cannot be used. The decision when to use them is a matter of judgment. When the problem is to compare relative amounts of change, in either a ratio or percentage, the ratio scale is the one to be preferred. If, on the other hand, the comparison is to be one of relative magnitudes only, the arithmetic ruling conveys the better visualization.

Ratio curves cannot be used for transition from plus to minus quantities as in the case of profits and losses. The reason for this is that zero or "0" cannot be indicated. The graduations can extend downwards in decimal divisions of 1 as 0.1, 0.01, 0.001, etc., but never to zero. Negative quantities, however, can be shown by themselves by inverting the sheet. (See Chapter XI, towards the end.)

CHAPTER VIII

THE TIME ELEMENT—HOW IT IS ARRANGED ON THE GRAPHIC CHART SHEET

To present a statistical fact upon a graphic chart sheet it is necessary to have a means of showing the relation of one measurement with respect to another. In business it is usually the relation of magnitude with respect to time.

It is now the generally accepted practice to indicate the periods of time along the horizontal base-line of the chart and to indicate the means for measuring the values of the magnitudes, or the "scale," as it is called, along the left vertical edge of the chart.

In the chapter, "How a Line Chart is Made," we referred to the time base-line and the vertical scale-line and explained briefly the purpose of each. We will now go into a little greater detail and show how the time base-line and scale should be marked.

In general, equal distances should represent equal facts. That is, on the time base-line equal distances should indicate equal periods of time; and on the scale, equal distances should represent equal differences in magnitude, if plain ruling, and equal differences in percentage, if ratio ruling.

It is desirable to have the graphic chart well balanced, i. e., neither high and narrow, nor low and wide. The former would result if the length covered by divisions on the time base-line were short as compared with the height of the vertical scale; the latter, if the length covered by "time" were long as compared with the height of the scale. Therefore, aim to have the lengths covered by the horizontal "time" divisions and by the vertical scale somewhere near equal.

Your data may be recorded in figures of hourly, daily, weekly, bi-monthly, monthly, quarterly, half-yearly, or yearly performance. The most commonly used are hourly, daily, weekly, monthly and yearly records, and these will be taken up in the following illustrations.

Hourly Records. Suppose, for example, your data are as follows:

TIME	PRICE
10 A. M.	50
11	48
12	46
1 P. M.	45½
2	47
3	48¼

you should use a graphic chart sheet that will give you the time element laid out with the hours evenly spaced. This is shown by Fig. 1.

Two or more days, recorded by hours, may be shown as in Fig. 2.

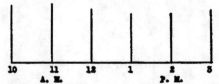

Fig. 1—Time Element of One Stock Market Day by Hours

Where the hourly records end for the day, the lines are accented to visualize the period of the day.

Daily Records. If your data are as follows:

Date	Sales
Jan. 2	$1,267.20
3	1,432.57
4	996.23
5	783.42
etc.,	

for one week, your time element should be arranged as shown in Fig. 3.

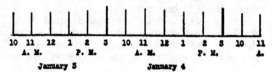

Fig. 2—Two or More Days Recorded by Hours

If the period is a full month instead of a week, the time element is indicated as shown in Fig. 4.

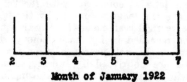

Fig. 3—One Week Recorded by Days

In like manner a daily record may be kept for any period of time. Fig. 5. shows the time element on a graphic chart sheet for a daily record over a year. This illustration, which is a small section of

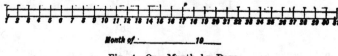

Fig. 4—One Month by Days

the full-sized sheet, shows the actual size of the ruling on the graphic chart sheets. Each month of the year is represented on the full sheet and each month is so divided that it has the correct number of divisions for its days. If the record is for a fiscal year instead of a calendar

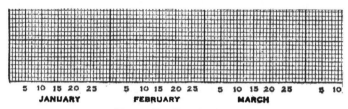

5   10   15   20   25   5   10   15   20   25   5   10   15   20   25   5   10

JANUARY   FEBRUARY   MARCH

Fig. 5—One Year by Days

year, the time element is arranged as shown in Fig. 6, where each month is divided into 31 divisions or days and is adaptable to one month as well as another.

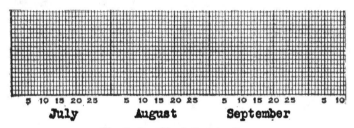

5   10   15   20   25   5   10   15   20   25   5   10   15   20   25   5   10

July   August   September

Fig. 6—One Fiscal Year by Days

Weekly Records. If your data are as follows:

WEEK OF		SALES
Jan. 2 to 7		$5,568.20
9	14	4,987.45
16	21	4,656.89
23	28	5,900.50

etc., use a ruling with the horizontal line divided into 52 parts, Fig. 7. The numbers 7, 4, 4, etc., are the dates of the first Saturday in each month of 1922. It is assumed that the weekly records end with Saturday and therefore each division represents a Saturday. This method is considered a good one for keeping weekly records. If the weeks are numbered 1, 2, 3, etc., up to 52, it would be difficult to identify the 36th week, for instance.

Fig. 8 shows another way in which the sheet may be arranged for a weekly record of a year. The average month consists of 4⅓ weeks and therefore vertical lines are drawn dividing the 52 divisions into 12 parts, or every 4⅓ divisions apart.

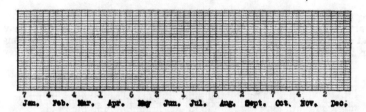

Fig. 7—Year of 1922; One Year by Weeks. First Saturday in Each Month
Indicated by 7, 4, 4, etc.

Monthly Records. If your data are as follows:

DATE	SALES
Jan. 1921	$29,387.46
Feb.	25,459.20
Mar.	27,222.17

etc., and cover the period of a year, the time element is arranged as

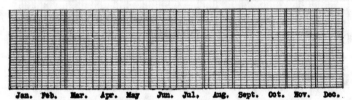

Fig. 8—One Year by Weeks. Each Month Containing 4⅓ Weeks

shown in Fig. 9.

Fig. 10 shows an arrangement for data covering a period of two years.

Year of 19____

Fig. 9—One Year by Months

This is excellent for comparing one year with the previous year.

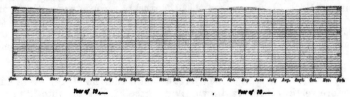

Year of 19,____   Year of 19____

Fig. 10—2 Years by Months

A chart for a monthly record over a period of five years is shown by Fig. 11.

If your data cover the period of several years, ten years for instance,

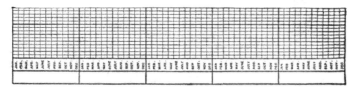

Fig. 11—5 Years by Months

the time element is spaced as shown in Fig. 12. Here every 12th division is accented in order that the years may stand out on the completed graphic chart.

Yearly Records. For yearly records over a period, each division on

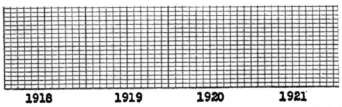

1918   1919   1920   1921

Fig. 12—Spacing of Time Element for Monthly Records over a Period of Years

the horizontal line is used to represent a year. The arrangement shown in Fig. 7 (called One Year by Weeks) can be used for a period of 52 years. It will then be known as 52 years by years. For 26 years every other division could be used; for 17 years, every third division; for 13 years, every fourth division; etc.

Codex Poly-Purpose Graphic Chart Sheet. The ruling illustrated in Fig. 13 is manufactured by the Codex Book Co., Inc., and is one of a set of three which they have named the Poly-Purpose Series. It has 72 equal divisions, horizontally. The 72 divisions may be marked off to

Fig. 13—Codex Poly-purpose Graphic Chart Sheet.

correspond with any desired periods of time—hours, days, weeks, months, etc.,—as will be seen from the illustrations. This ruling may be used for:

Records by seconds covering 1 minute.

Records by minutes covering 1 hour.
Hourly records covering 3 days.
12-hour shift records (2 per day) covering 36 days.
8-hour shift records (3 per day) covering 24 days.

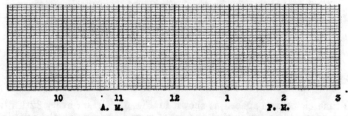

Fig. 14—Codex Poly-purpose Graphic Chart Sheet Arranged for One Day by Hours. This arrangement may also be used for a Week by Days

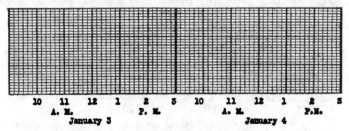

Fig. 15—Codex Poly-purpose Graphic Chart Sheet Arranged for Two Days by Hours

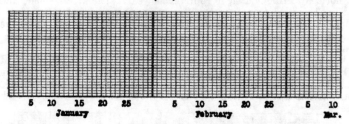

Fig. 16—Codex Poly-purpose Graphic Chart Sheet Arranged for Two Months, Plus, by Days

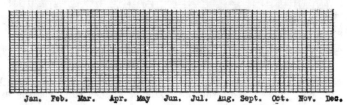

Fig. 17—Codex Poly-purpose Graphic Chart Sheet Arranged for One Year by Months

Full-week daily records (7 days) covering 2 months or 10 weeks.

Workday-week daily records (6 days) covering 12 weeks.
Weekly records covering 1 year.
Monthly records covering 6 years.
Yearly records covering 72 years.

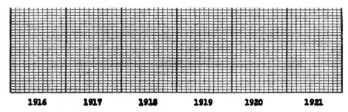

Fig. 18—Codex Poly-purpose Graphic Chart Sheet Arranged for Six Years by Months

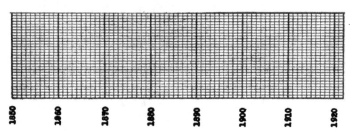

Fig. 19—Codex Poly-purpose Graphic Chart Sheet Arranged for Seventy-two Years by Years

CHAPTER IX

How to Assign the Scale to the Plain Ruling

The scale, so far as the graphic chart for business is concerned, is a straight line on which are marked off a series of spaces, graduated and numbered, for measuring the magnitudes or values of the points plotted on the chart sheet. The scale is usually numbered on the left vertical

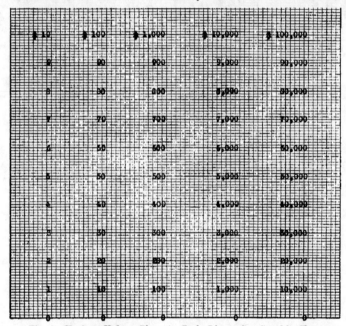

Fig. 1—Various Values Given to Scale Lines of a Graphic Chart

edge of the chart sheet, if only one scale is shown. Frequently, however, for clearness a scale is shown on both the left and right edges.

The ruled sheets which are sold today for making business charts are already spaced off and horizontal lines are drawn across the sheet from one edge to the other. The spacing of these lines varies, but the most useful for the business man is a sheet with lines which are far enough apart to make plotting easy and which are decimally divided; that is, with accented lines every five, ten or twenty spaces apart.

It only remains, then, to assign suitable values to these divisions and

38

the scale is complete. In doing this there are three things to remember.

1. The scale on chart sheets with plain ruling should always begin with zero. Charts on plain ruling are frequently seen with scales which do not begin with zero, but such charts are not correctly drawn and are misleading. The function of the chart with plain ruling is to give a picture of comparative numerical differences. To obtain a true picture, the common basis of comparison, which is the starting point, zero, should be shown.

2. Number the scale uniformly. That is, as you progress up the scale from zero, let equal distances on the scale represent equal numerical differences. For example, if the scale were divided correctly into half-inch spaces, begin at the bottom with zero and number each half-inch division point so that the same numerical difference exists between the successive points; as 0, 10, 20, 30, 40, etc., or 0, 25, 50, 75, 100, etc., or 0, 1000, 2000, 3000, etc. See Fig. 1.

In using the purchased ruled sheets it is best to so number the scale that its units shall be multiples of the number of divisions into which the paper is ruled. This greatly facilitates interpolating the intervening values. For example, in plotting dollars to units of $100, it would be better to use a sheet divided into twenty spaces between the main division points, where each space would equal $5., than one divided into six spaces, where each division would equal $16.67.

3. Number the scale in such a way that the range covered by its full height will just about cover the range of the data to be plotted. In other words, if your data vary from $25 to $3000, let the full length of your scale (or as much of it as is practicable) cover a range of $0 to $3000, not $0 to $50,000, or some other value considerably greater than $3000. You will then have a better balanced chart and the values of the points on the curve may be read with greater accuracy. Fig. 2 shows a well proportioned graphic chart.

Let us again consider the receipts of our small business. They were:

Monday	$350.
Tuesday	300.
Wednesday	400.
Thursday	450.
Saturday	450.
Friday	500.

The greatest amount is $500 and therefore we must make our scale include this figure. The range of our data is from, the minimum, $300, to $500, the maximum.

In applying to our graphic chart sheet a scale to cover this range we must remember that zero should be shown. Therefore our scale will range from 0 to $500. In Fig. 1 none of the scales is just what we want, but the second, 0—$100, if multiplied by five, would give us the range we require. This range, 0—$500, would take in the full length of the scale and give a well proportioned chart, Fig. 3. There are 10 subdivisions to each main scale division in Fig. 3. Therefore, as the main divisions have a numerical difference of $50, each subdivision will have a value of $5. It is not necessary to mark each $5 line with its

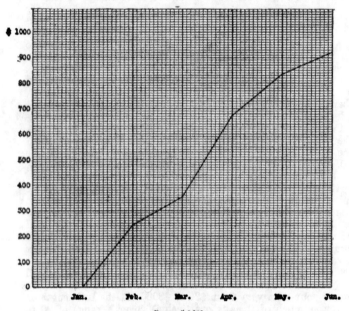

Year of 1921
Fig. 2—A Graphic Chart Properly Proportioned

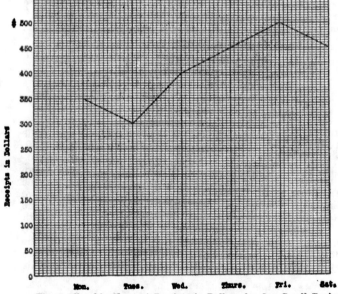

Fig. 3—Graphic Chart of Receipts in Dollars for Our Small Business

value, as it is easy to read the intermediate values when each $50 point is indicated on the scale.

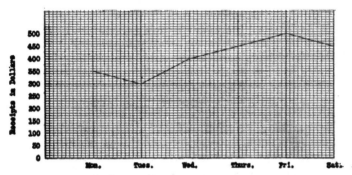

Fig. 4—Scale Applied to 50 Vertical Divisions

If the ruling of our graphic chart sheet is such that the scale has 50 divisions, each division line is equal to $500 divided by 50, or $10. Fig. 4 shows how the scale is applied to a ruling of this sort.

As our numerical values are very seldom in even $50 or $100 amounts, it is frequently necessary to interpolate. For instance, suppose that on Tuesday our receipts were $303.22 instead of $300.

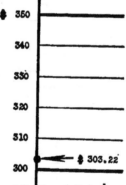

Fig. 5.—Enlarged Portion of Scale from $300 to $350

On Fig. 4 each line represents $10. The amount $303.22 would, therefore, be about one third the space between the $300 scale line and the next line above it, $310. Fig. 5 shows the portion of the scale from $300 to $350 greatly enlarged. The point, $303.22 is plotted about one third the distance between $300 and $310.

If our daily receipts were as follows:

Monday	$345.76
Tuesday	303.22

Wednesday	389.50
Thursday	458.17
Friday	506.25
Saturday	444.10

we could use the same scale as shown in Fig. 4, but the numerical values would be interpolated as shown in the enlarged section of the scale from $300 to $500, Fig. 6.

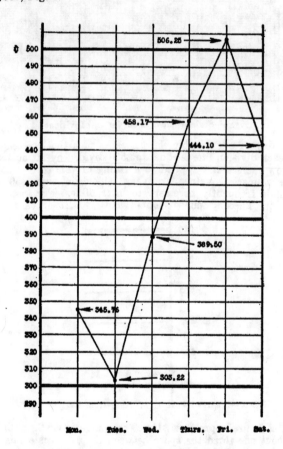

Fig. 6—Interpolated Points on Enlarged Section of Scale

Of course, it is not possible to read the cents from the chart, and the units place of the dollars is estimated, but the estimating may be quite accurate when one becomes accustomed to it.

'As heretofore stated, the scale for charts plotted on plain ruling should start with zero. If the curve varies from positive to negative values (as from profits to losses), the zero line should be drawn to allow proper space for the positive values above, and the negative values below, the zero line. This is illustrated by Fig. 7.

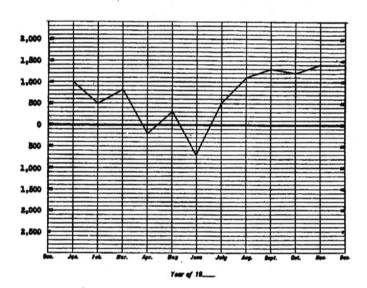

Fig. 7—Curve Running to Negative Values

If the chart is to be a cumulative one, i. e., if it will be added to from day to day, or from month to month, etc., a proper allowance must be made for the larger quantities that will occur in the future.

Fig. 8 shows a graphic chart that has been plotted to August and that is to be continued for succeeding months. The highest point to August on the chart is $8,400. The scale on this chart will allow a maximum of $10,800 which is higher than it will probably go, based on past experience.

It should be kept in mind at all times that clearness is the essence of the picture shown on the graphic chart sheet, and every effort should be made to achieve that end.

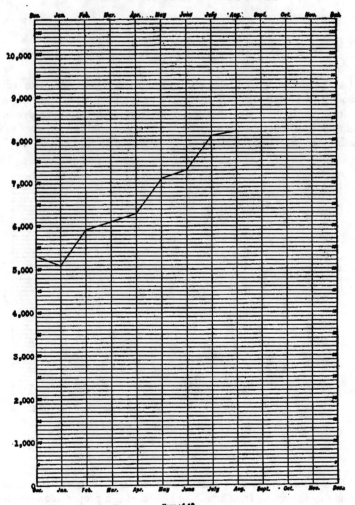

Year of 19_____
Fig. 8—Allowing For Future Values on a Cumulative Chart

CHAPTER X

How to Apply the Proper Scale to the Ratio Ruling

Upon most of the chart sheets with ratio ruling, which are sold to-day, guide figures are already printed as aids in marking the scale correctly. See the small figures along the right edge of Fig. 1.

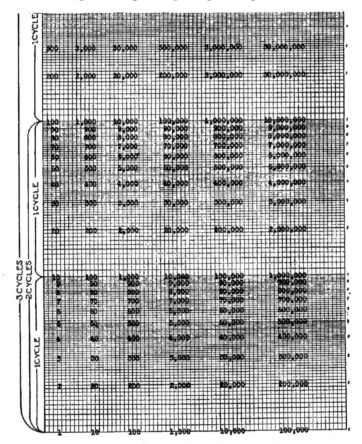

Fig. 1—How the Scale on the Ratio Chart is Marked

It will be noted that these figures begin at the bottom with 1, run up through 9 and then start over again with 1. Each complete set, 1 through 9, is called a "cycle." The numbers on the different cycles are the same because, as mentioned before, they serve merely as guides, and they may be followed (or preceded) by the necessary number of zeros to give the scale the range that is desired.

Ratio ruling is mathematically determined and the relation existing between the various divisions never varies, no matter what the size of the cycle may be. In other words, whether we have a three-inch cycle or a twenty-inch cycle, the same relation will exist between 1 and 2, 2 and 3, 3 and 4, etc., on the one as on the other.

Different from the plain ruling, the ratio ruling has no zero. The reason for this is explained at greater length later. The bottom cycle may begin with any multiple of 10, such as 0.001, 0.01, 0.1, 1, 10, 100, 1000, etc. Whichever multiple of 10 it begins with, the cycle above must begin with the next higher multiple of 10, and so on. For example, if you have a three-cycle chart sheet, you may begin it with any multiple of 10 you desire. If you begin with 1 the scale progresses 2, 3, 4, etc., up to 10; then 20, 30, 40, etc., up to 100; then 200, 300, 400, etc., up to 1000. If you were to begin with 0.01 the scale would run 0.01, 0.02, 0.03, 0.04 up to 0.1; then 0.2, 0.3, 0.4, etc., to 1; then 2, 3, 4, etc., to 10. If you were to begin with 10,000, the scale would be numbered 10,000, 20,000, 30,000, 40,000, etc., to 100,000; then 200,000, 300,000, 400,000, 500,000, 600,000 etc., up to 1,000,000; then 2,000,000, 3,000,000 etc., up to 10,000,000. Thus, note that the cycles may be so chosen that their number and scale may be made to embrace any range desired. Fig. 1 shows clearly how the cycles should be numbered. The reason for this is plain when it is remembered that on the ratio scale the same vertical distance always represents the same percentage difference. There is the same percentage difference between 1 and 2, 10 and 20, 100 and 200, etc., therefore the distance between them on the scale is the same. The same thing applies to any numbers which bear the same percentage relation to one another.

The values to apply to the lines of the ruling are determined by an inspection of the data. If your total sales for the period covered by the data ranged from $35,000 to $78,000, (high and low limits), your resulting curve would be within the limits of one cycle, which would be numbered from $10,000 to $100,000. Where the small guide figure, 1, appears at the bottom of the first cycle you would start your scale with $10,000. The scale would then run $20,000, $30,000, etc., up to $100,000. This is illustrated by Fig. 2.

If your total sales, or whatnot, for the period covered by the data, ranged from $35,000 to $134,000, your resulting curve would require two cycles. Your scale would then start at the guide number 1 (Fig. 3) with a value of $10,000, run up to $100,000 (1 on the next cycle) and continue up to $1,000,000.

Briefly, you should use one cycle if your data range from 0.01 to 0.1; 0.1 to 1; 1 to 10; 10 to 100; 100 to 1,000; 1,000 to 10,000, etc. In each case you will note that the figure at the top of the cycle is ten times that at the beginning, or bottom, of the cycle.

You should use two cycles if your data range from 0.01 to 1; 1 to 100; 10 to 1,000; 100 to 10,000, etc. In each case the figure at the top

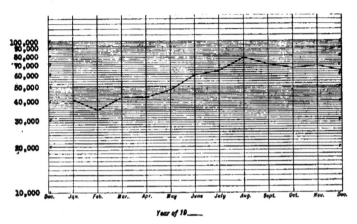

Fig. 2—Data Covered by a Single Cycle

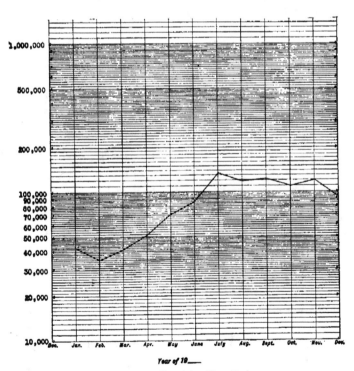

Fig. 3—Data Requiring Two Cycles

of the second cycle is one hundred times that at the bottom of the first cycle.

If your data range from 0.001 to 1; 1 to 1,000; 10 to 10,000; 100 to 100,000, you should use a ratio ruling of three cycles, noting that the

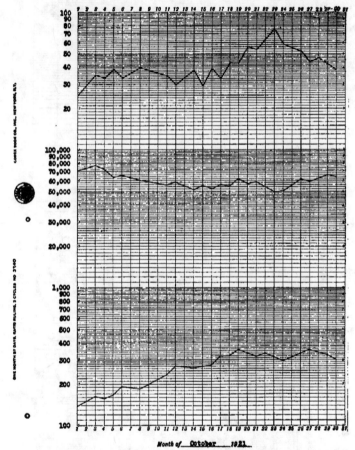

Fig. 4—Three Curves on a Single Graphic Chart Sheet
Plotted to Different Scales

figure at the top of the third cycle is one thousand times that at the beginning of the first cycle.

In general, then, a cycle of ratio ruling should always begin with some multiple of 10, and the value at the top, or end, of the cycle is always ten times that at the beginning, or bottom, of the cycle.

Therefore, as you progress up a scale consisting of several cycles of

ratio ruling, the value at the top of the first cycle is ten times that at
the bottom.   The value at the top of the second cycle is ten times that
at the bottom of the second cycle, and one hundred times that at the
bottom of the first cycle.   The value at the top of the third is ten times
that at the bottom of the third; one hundred times that at the bottom
of the second, and one thousand times that at the bottom the first cycle;
and so on.

The scale covering two or more cycles need not be continuous, how-

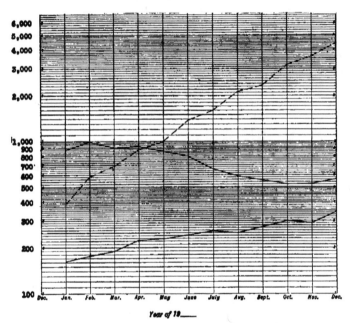

Fig. 5—Three Curves Plotted to the Same Scale

ever.   If desired, each cycle on the sheet may be scaled to cover a different range, and separate sets of data may be plotted, each according to
its own scale.

The curves may be compared with one another just the same as though
all were plotted to the same scale, for, on ratio ruling, so long as the
cycles are of the same size, the same percentage relations hold, no matter what the scale may be.   Fig. 4 shows three curves on the same sheet,
each drawn to a different scale.   Fig. 5 shows three curves plotted to
the same scale.

CHAPTER XI

How Ratio Charts May be Used

In the preceding chapters mention has been made here and there of many of the unique peculiarities of the ratio ruling. At the expense of seeming to repeat we are going to emphasize some of these peculiarities because they are at present not appreciated by the majority and yet they are the things which make ratio charts so valuable to the business executive.

What the Slope of the Curve Indicates. Where data with changing magnitudes are plotted on a ratio chart the steepness or slope of the lines is an exact measure of the rate of change in the data. This applies to the general trend of the curve as a whole or to the various portions which go to make up the curve. The following tells how the slopes may be interpreted:

1. If the curve is horizontal, neither ascending nor descending, the magnitudes are uniform, neither increasing nor decreasing. (1) of Fig. 1.

2. If the curve ascends and is straight, or very nearly so, the magnitudes are increasing at a uniform, or very nearly uniform, rate. (2) of Fig. 1.

3. If the curve ascends and bends upward, concave fashion, the magnitudes are increasing at an increasing rate. (3) of Fig. 1.

4. If the curve ascends and bends downward, convex fashion, the magnitudes are increasing at a decreasing rate. (4) of Fig. 1.

5. If the curve descends and is straight, or very nearly so, the magnitudes are decreasing at a uniform, or very nearly uniform, rate. (5) of Fig. 1.

6. If the curve descends and bends upward, concave fashion, the magnitudes are decreasing at a decreasing rate. (6) of Fig. 1.

7. If the curve descends and bends downward, convex fashion, the magnitudes are decreasing at an increasing rate. (7) of Fig. 1.

8. If the direction of the curve in one portion is the same as that in another portion, the percentage rate of change is the same for both portions.

9. If the curve is steeper in one portion than in another portion the rate of change is more rapid in the former than in the latter.

10. If two curves on the same ratio chart are parallel to each other the same rate of change exists in both.

11. If one curve is steeper than another the rate of change in the former is more rapid than in the latter.

12. A straight line, either actual or imaginary, drawn so as to represent the general trend of a curve will show by its slope the average rate of increase or decrease of the curve; and the deviations from this average may be readily perceived.

50

13. A straight line, either actual or imaginary, may be drawn between any two points on the curve, between which it is desired to know the rate of change, and the slope of this line will show the rate of increase or decrease between the two points.

Measuring the Percentages of Increase or Decrease on a Ratio Chart.
As we have said several times before, it is not possible to obtain a correct picture of relative rates of increase or decrease on plain ruling. One method of attempting to meet this shortcoming is to compute the ratios

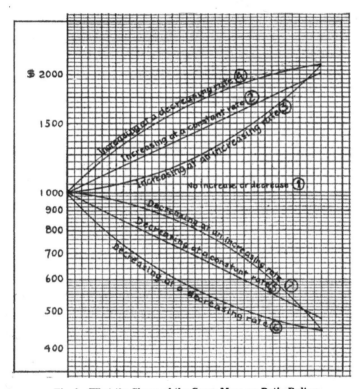

Fig. 1—What the Slopes of the Curve Mean on Ratio Ruling

desired and plot them on a separate chart. For example, if a chart had been made on plain ruling showing the actual figures of income and expense by months for one year, and it was then desired to know how the ratio of expense to income had varied throughout the year, it would be necessary to compute the percentage of expense to income for each month and then to plot these results on a separate chart. It is readily seen that this involves considerable calculation as well as the drawing of a new chart.

If the actual figures were charted on ratio ruling, not only would we obtain a correct picture of the relative rates of increase and decrease, but also we could read the amounts of the percentage of increase or decrease directly from the chart. The actual percentages of increase or decrease may be measured by the eye or they may be measured with a scale.

Remember that on a ratio chart a given vertical distance on the scale always represents the same percentage difference, no matter where it is on the chart. Therefore, in figuring the percentage difference between any two points, always take the vertical distance between the points and not the distance along the curve connecting them. This is illustrated by Fig. 2. The percentage difference between the points A and B is measured by the vertical distance, B C, between the points and not by the distance A B.

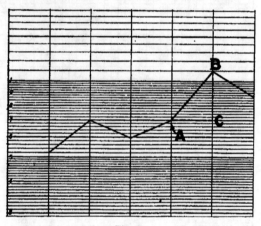

Fig. 2

In using a ratio chart of a given scale, the eye soon becomes trained so that it can figure out with considerable accuracy the percentage differences with respect to that scale. That is, a certain vertical height on the scale will represent a percentage difference of 10%, another height a percentage difference of 100%, etc. Of course, on different scales the same percentage would be represented by different heights. For example, 10% would be represented by a certain vertical height on a three-inch cycle, but on a nine-inch cycle the vertical height, for 10% would be considerably greater.

Where more accurate measuring of the percentages of increase and decrease is desired than can be obtained with the eye, this may be accomplished by the simple expedient of making a percentage scale to fit the chart on which the curves are drawn. That is, when using a chart with three-inch cycles of ratio ruling, use the three-inch cycles for the percentage scale; when plotting on a chart with nine-inch cycles of ratio ruling use nine-inch cycles for the percentage scale, etc.

To illustrate the construction and use of the percentage scales we will take as an example a ratio chart with two cycles of ratio ruling upon which a series of curves have been plotted. The values range between 10 and 1,000, Fig. 3. We wish to construct a percentage scale that will enable us to measure the percentage of increase or decrease between any two points on the chart. Let the attention rest upon the middle line of the two cycles, M, Fig. 3. The scale value given to this line is 100. The value of the next main division above is 200. 200 is twice 100, or 100% greater. The value of the next division above is 300. 300 is three times 100, or 200% greater. And so on up the scale. Thus the even hundred-per cent. increases are determined from 0% at the 100 scale point to 900% at the 1000 scale point. Intermediate division points may be determined in the same manner. The value of the first sub-

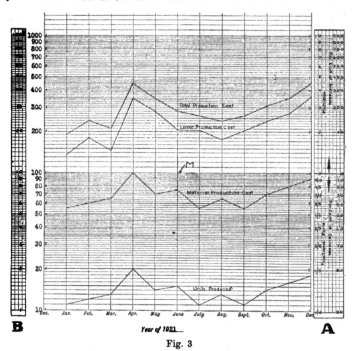

Fig. 3

division above 100 is 110. 110 is ⅒ or 10% greater than 100. So on up the scale. By interpolating the values between the 10% points it is possible to read such a scale to the nearest 1% with a tolerable degree of accuracy, up to about 300 % or 400% and to the nearest 5% from 400% up. This accuracy is almost always sufficient for executive control.

Therefore, we may construct a scale of multiples, and percentages of increase, see A, Fig. 3. This is most easily done by taking a narrow strip of the same paper as that upon which the curves are drawn, and making it as shown in the figure. It is handiest to take a strip off either the left or right-hand border of the sheet, as they are usually already marked with scale values, and it is only necessary to affix the corresponding percentage values opposite the numbers on the strip.

For a scale to measure the percentages of decrease, once more consider the 100 line, M. The next main division below this is marked 90. 90 is $\frac{9}{10}$ or 90% of 100. The value of the next main division bebelow is 80. 80 is $\frac{4}{5}$ or 80% of 100, or 20% less than 100, and so on down the scale. Intermediate points may be determined as explained above. Therefore, we may construct a scale of fractional parts and percentages of decrease, see scale A, Fig. 3. We will use the same strip of paper and number it as shown in the figure.

We now have a means for measuring the percentages of increase or decrease between any points on the curves.

We wish to find out, for example, what the percentage of increase was, in Total Production Cost, for April over March. For this we use the upper part of the scale marked Per cent. of Increase. Place the right edge of the scale along the March line, with its 0 point coinciding with the point denoting the March value of the Total Production Cost, Fig. 4. Then project a horizontal line from the point denoting the April value of Total Production Cost to the scale, and we read 114%. The percentage is measured by the vertical distance between the points. Fig. 4 shows the section of Fig. 3 enlarged.

The same method may be used to determine the percentage of increase between any two points on any one of the various curves, or between any two points, one of which is on one curve and one on another curve. If the percentage scale is not long enough it may be extended to give any desired range.

To ascertain the percentage of decrease in Material Production Cost between April and May, we will use the lower part of the scale, marked Per cent. of Decrease. Place the right edge of the scale along the April line, with its 0 point coinciding with the point denoting the April value of Material Production Cost. Then project a horizontal line from the point denoting the May value of Material Production Cost to the scale and read 30%. (See Fig. 3.)

The same method may be used to determine the percentage of decrease between any two points, either when both points are on the same curve or when one point is on one curve and one on another curve. The range of this scale may be extended to meet any requirement.

A Scale for Determining Component Parts. Frequently we plot on one chart several curves, where one curve represents the sum of the others. That is, we might have on a chart curves for Total Operating Expense, and others for the miscellaneous expenses, the sum of which would equal Total Operating Expense. Or, we might show Income as the total of Manufacturing Cost, Sales Cost, Overhead and Profit. We can construct a scale by means of which we may read off directly the percentage which each of the component parts bears to the total. If a line, say the 1,000 line, Fig. 3, is taken as a base of 100%, then the 900 line would be 900/1000ths, or 90%; the 800 line would be 80%; and so on down the scale as far as it is necessary to go. Therefore, to construct the scale take a narrow strip of the chart paper, similar to that used for the percentage scale. Starting at the top with 100%, mark off the main divisions—90%, 80%, 70%, etc., down to the bottom. The intermediate lines of the scale make it possible to read to the nearest 1% from 100% down to 50%; to 0.1% from 50% down to 5%; and to 0.01% from 5% on down. At the left side of Fig. 3 we have shown such a scale with a range from 1% to 100%. See scale B, Fig. 3.

In Fig. 3 the sum of the Labor Production Cost and Material Production Cost equals the Total Production Cost. Let us suppose that it is desired to know what percentage each of the former bears to the total for June. Place the edge of the Component Parts Scale along the June line so that the 100% point on the scale coincides with the Total Production Cost point for June. Then we go down the scale until we come to the Labor Production Cost curve, which intersects

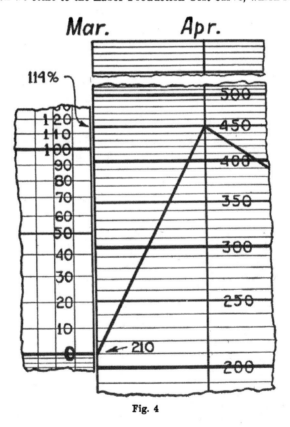

Fig. 4

the scale at 73%; then we continue down the scale until the Material Production Cost curve intersects it and we read 27%. The sum of 73% and 27% is 100%, which shows us that our scale readings are correct. No matter how many curves there are on the sheet the same method applies. It is readily seen that it furnishes a means of quickly and easily checking up many facts vital to economic management, such, for example, as percentage of orders received to quota, and, in fact, the percentage of any actual performance to a set standard; percentage of collections to receivables; percentage of gross sales to advertising expenses; etc.

If the sum of the various components reaches a total which is considerably larger than any one of the components, it may seem preferable, in charting the data, to change the chart scale for the total, rather than to use a chart sheet of greater range, upon which the total could be plotted to the same chart scale as the components. If the chart scale is changed

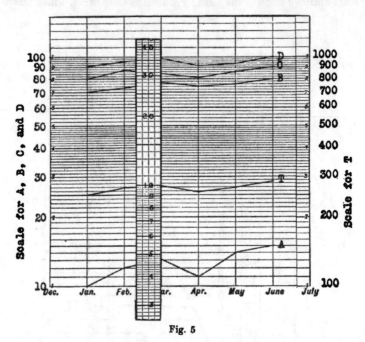

Fig. 5

the Component Parts Scale may be used just the same, but in a little different way. For example, see Fig. 5. Here we have four components A, B, C, D, all of which have values such that they can be charted upon one cycle of ratio ruling. Their total, however, is of such a magnitude as to require two cycles of ratio ruling, plotted to the same scale. Therefore, instead of using a two-cycle paper, we will change the scale so that the range, instead of being 10 to 100 as it was for the components, will be 100 to 1,000. To measure the percentage of the component parts—A, B, C, D—to the total T, for the month of March, place the edge of the Component Parts Scale along the March line so that the 10% point on the scale (not the 100% point, as in the ordinary case) coincides with the T curve, Fig. 5. Then we go up the Component Parts Scale and read off directly the valves of B, C, and D at the points where the curves intersect. These percentages are 28, 31 and 36. Then for the A curve we go down the Component Parts Scale and where the A curve intersects we read 5%. The sum of 28, 31, 36, and 5 is 100, as it should be.

Shifting the Curves on the Ratio Charts. We have just shown in the last paragraph how it is possible to plot curves on the same cycle to different scales. The same principle applies to any number of cycles, and, frequently, by changing the scale, a two or three cycle chart sheet may be utilized for data, which, if plotted to the same scale, would require 5 or 6 cycles. Futhermore, the correct determination of percentage differences is in no way affected by this scale changing, for

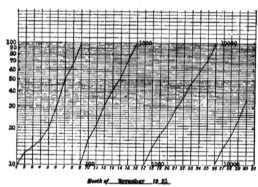

Month of   November   19 21

Fig. 6

the feature which is peculiar to the ratio chart—that the same vertical distance on the chart always represents the same percentage difference —holds true no matter what the scale is, or upon what part of the sheet the curves are. From this follows a fact to be remembered about the plain chart. On the plain chart the zero base-line should always be shown, as the function of the chart is to picture relative magnitudes, and the basis of such a comparision should always be the zero base-line. On the ratio chart no specified base-line need be shown. In fact, any value on the ratio scale may be taken as the base-line. It is customary, but not necessary, that the bottom line should be some power of 10.

Another way the feature of changing scales may be utilized is where a certain scale range has been assigned to a cycle or cycles and then it is found that the data exceed this range, such, for example, as in a cumulative curve. The cycle may simply have a new scale value assigned to it and the curve may start anew from the point where it ran off the first scale. Fig. 6.

For comparing any two curves on ratio ruling, the curves may be moved up and down at will, in order to bring them closer together for purposes of comparison. This may be accomplished by changing the scales on the cycles of ratio ruling, as explained above, or by drawing curves on separate sheets of transparent paper and then placing one sheet above the other and moving it about until the curves are in the position desired.

How to Multiply on Ratio Ruling. Figure 7 shows a series of curves representing units of production and cost of production. Let us assume, for an example, that we have only two curves, A and B, Number of Units Produced and Cost per Unit Produced. From these we wish to determine the Total Production Cost, for the various months, January to December. For illustration, we will take the month of May.

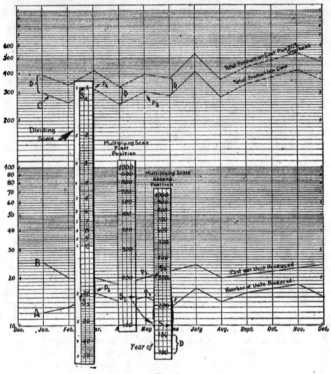

Fig. 7

We wish to multiply the quantity represented by P_1, by the quantity represented by P_2. For a multiplying scale we use a narrow strip made from one or more cycles of ratio ruling. These cycles must be the same size as those upon which the curves are plotted.

For numbering the multiplying scale, begin it with the number which is the square of the number which begins the chart scale. (A "square" is the product of a number multiplied by itself.) That is, if the first cycle on the chart begins with 10, begin the multiplying scale with 10 × 10, or 100. See Fig. 7. If the chart scale starts with 1, begin the multiplying scale with 1×1, or 1. If the chart scale starts with 100, begin the multiplying scale with 100×100, or 10,000, and so on. To multiply P_1 and P_2 we make the bottom of the multiplying scale coincide with the base-line of the chart, as in the First Position, Fig. 7. This point we will call S_1. Next we lower the multiplying scale until S_1 coincides with the base-line of the chart as shown in the Second Position, Fig. 7. Finally we mark off on the multiplying scale a point, S_2, opposite P_2. S_2 is the answer. In our illustration, the value of P_1 and P_2 are 15 and 20. $15 \times 20 = 300$, which is what the point S_2 on the multiplying scale reads. Similar multiplications may be made for the other months. The values obtained may be charted as fast as they are determined, see P_3, Fig. 7. Then the curve connecting these points will be that of Total Production Cost.

Suppose, now, we should like to know what our Total Production Cost plus overhead would be. Assume an overhead of 30 per cent. Then vertically above the points on the curve of Total Production Cost we place points at a distance representing 30 per cent. on the scale. The bottom figure on the multiplying scale is 100. 30 per cent. more than this would be 130. Therefore the distance D, (see Second Position, Fig. 7.) between 100 and 130, is what we add above the various points of the Total Production Cost curve to get the points for the curve of Total Production Cost plus 30 per cent. Overhead. Connect these points and we have the curve as shown. This, of course, runs parallel to the curve of Total Production Cost.

How to Divide on Ratio Ruling. Let us assume, now, that we have only the two curves A and C, Number of Units Produced and Total Production Cost, Fig. 7. From these we wish to determine the Cost per Unit Produced, for the various months. For illustration we will take the month of March. We wish to divide the quantity represented by the point P_4 by the quantity represented by the point P_5. For a dividing scale we use a narrow strip made from one or more cycles of the same size as those upon which the curves are plotted. But for dividing, the scale is reversed. That is, the ruling is turned end for end; so that instead of beginning with 1 at the bottom and progressing upward, we begin with 1 at the top and progress downward on the scale. See Dividing Scale, Fig. 7. To divide P_4 by P_5 we make the top of the dividing scale, S_4 coincide with P_4, the edge of the scale being vertical. Then where the point P_5 hits the dividing scale we get our answer S_5. In our illustration the values of P_4 and P_5 are 320 and 16. $320 \div 16 = 20$, which is what the scale reads at S_5, where the point P_5 meets it.

The Ratio Ruling has no Zero. It has already been remarked several times that the ratio ruling has no zero base-line. The function of the ratio chart is to compare relative percentages and there must necessarily be some amount to compare with. It does not matter how small the amount is, but it cannot equal zero. We could not say, for example, that a magnitude is 50 per cent. greater than zero. In other words, as long as there are magnitudes to be plotted, no matter how small they may be, the per cent. of decrease can never equal 100 and therefore can never reach the zero point. Extending a ratio scale to zero would be equivalent to reducing a finite number to zero by successive divisions, and neither can be accomplished short of infinity.

The fact that the ratio chart has no zero has a certain advantage in that it focuses the attention upon the ratio of one magnitude to another and their rates of change, rather than upon a comparison of their numerical differences.

It is possible to show a zero line on a ratio chart by drawing a line at any convenient distance below, and parallel to, the base-line of the chart and arbitrarily calling it zero. It must be remembered, however, that the function of the ratio chart—the correct measuring of percentage differences—will cease to operate between any two points, one of which has been plotted on the arbitrarily chosen zero-line.

Minus quantities may be shown on the ratio chart by turning the sheet upside down. If the magnitudes vary from plus to minus, thus

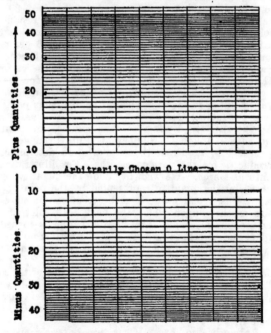

Fig. 8—Showing Minus Quantities and Zero on Ratio Chart

passing through zero, a zero line may be shown as explained above, see Fig. 8. Once more, however, be reminded that it is not possible to measure the percentages of increase or decrease correctly between any two points if one of them is on the zero line.

Forecasting by Means of the Ratio Chart. Every business man likes to look into the future and predict what his income, expenses, production, etc., will be next month, next year, or five years hence. This may be merely a matter of interest or it may enable him to make a budget or to lay down standards of performance.

As an aid to such forecasting, graphic charts are very valuable. As we have remarked elsewhere, a graphic chart does not, of itself, furnish a means of knowing just what will take place at some future time. It does furnish a basis whereby past events may be taken as a criterion of what is likely to transpire in the future. But in connection with the use of graphic charts, the same discretion and judgment must be exercised as in everything else.

Usually, in business, we make our forecast by assuming a uniform percentage of growth. We predict, for example, that our income will increase at the rate of 20 per cent. a year. Uniform percentage of growth is shown by a straight line on the ratio chart. Usually, to make a forecast on a ratio chart, all we need to do is to extend the line representing the rate of increase or decrease in the past. This extended line will show us what we may expect in the future, if the same rate of increase or decrease obtains as in the past.

At present the graphic chart with plain ruling is generally used for forecasting and the conclusions drawn therefrom are very apt to be incorrect. For where it is intended to show uniform percentage of change, a straight line on plain ruling does not show such a result. A straight line on plain ruling shows an increase at a decreasing rate. Increase at a uniform rate would be a curve, concave, and upward. Those not very familiar with charts are prone to interpret the straight line on plain ruling as indicating a uniform rate of increase and to interpret the true curve of uniform increase as indicating that the rate of growth is increasing. Therefore, it is advisable to use ratio rulings for charting all data where forecasting is contemplated.

Charting Price Fluctuations of Stocks and Bonds on Ratio Ruling. It is becoming more and more the custom for business men to keep graphic charts of the price fluctuations of stocks and bonds. These graphic records may be extended over a very long period in order to get as complete a history as possible, or they may extend over a comparatively short interval where it is desired to know the fluctuations with respect to a certain amount, as, for example, a known average, or with respect to the purchase price, etc.

This applies not only to the bankers and brokers whose business it is to deal in these things but to every one who has his money invested in them.

Because this is so we deem it worth while to call attention to the particular value of ratio ruling in this connection.

In considering various stocks and bonds, if we wish to buy or sell at a favorable high or low point, we should be concerned not so much with the number of points rise or fall as with the *percentage* of rise or fall. In other words, our return will be greater on a five point rise from a stock for which we have paid 75 than from one for which we have paid 250. In the first case it will be $5/75$, or 6.7 per cent.; in the second, $5/250$, or 2 per cent.

On the plain ruling a five point rise would show up the same on any part of the chart; that is a rise from 75 to 80 would be of the same vertical height on the plain ruling as from 250 to 255. On ratio ruling, however, the five point rise from 75 to 80 would be shown by a vertical distance considerably greater than the five point rise from 250 to 255.

On the plain ruling both five point rises would appear to be of equal value; on the ratio ruling they show up in their true relation, and could be measured by the scale illustrated in Fig. 3 of Chapter XI.

CHAPTER XII

Hints in Making Line Charts

The following are hints directly applying to the Line Chart. General hints and cautions in making graphic charts for business purposes are given in Chapter XIX.

Distinguishing One Curve from Another. At the time a chart is drawn explanatory notes should be used unsparingly and all curves or diagrams should be so keyed or lettered that there will be no question as to just what each represents when the chart is referred to later. When it is not necessary to make blue-prints, different colored inks or water-color paints may be used in a drawing pen to distinguish one curve from another and a key showing what each color represents put on the sheet, as for example:

Gross earnings: ——— (red line)
Expenses: ——— (blue line)
Net revenue: ——— (green line)

For ordinary graphic chart work, when there are but one or two

Fig. 1—Bottle of
Drawing Ink

curves to be drawn on a single sheet, the best practice is to draw a curve first in pencil, and then, being sure that it is correct, go over it with

Fig. 2—A Drawing Pen

black india ink. Fig. 1. illustrates the usual type of bottle in which india ink comes.

A drawing pen is illustrated in Fig 2. With this type of pen uniform lines of various widths may be drawn.

63

For blueprints, the most legible way of distinguishing one curve from another, is by drawing one curve with a solid line, another with a line

Fig. 3—Solid and Broken Lines to Distinguish Various Curves

made up of dashes, another of dots and dashes, etc. This is illustrated by Fig. 3.

Identifying the Curves on the chart may be done by either using a key, as, for instance, a solid line to represent gross earnings; broken line, expenses; etc., or by lettering the information on the curves themselves, or

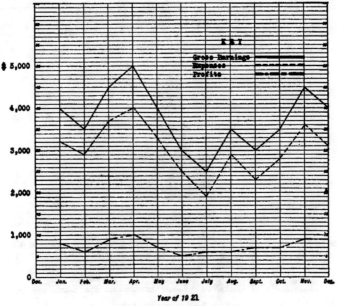

Fig. 4—Curves Identified by Key

both. If a key is used, it should be, as far as possible, standard for all the data charted by the company. Fig. 4 shows how a key may be arranged on a graphic chart sheet. Some prefer, however, to letter the curves directly, as shown in Fig. 5.

Solid lines may be drawn for each curve on a graphic chart sheet where the curves do not cross one another. When the curves cross, it is

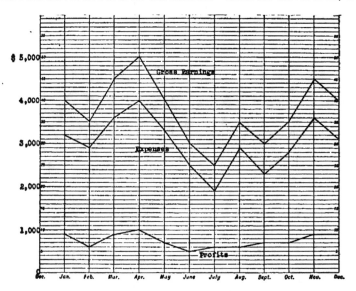

Fig. 5—Curves Identified by Lettering Directly

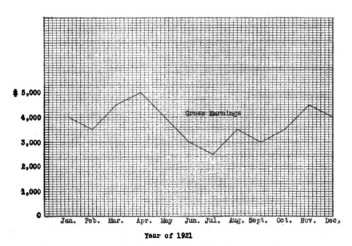

Year of 1921

Fig. 6—A Sheet with More Vertical Lines than are Necessary

preferable to show them either with broken lines of various combinations of dots and dashes or with different colored inks.

Simple Ruling Desirable. Too many cross-sectional lines ruled on a graphic chart sheet may tend to confuse. Where amounts are to be read from the chart, it is important to have enough intermediate lines so that little interpolation is necessary. Where it is simply a case of noting and studying the trend of a curve, the fewer lines, the better. This is brought out by an inspection of Figs. 6 and 7.

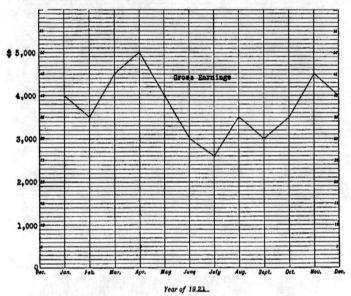

Fig. 7—A Sheet with Only the Necessary Lines

Pasting Graphic Chart Sheets Together. If it is necessary to make a chart to cover a period of time too long for one sheet of paper, or if the scale required is beyond the range of one sheet, the range may be extended by pasting together two or more sheets. Use a good quality of paste spread very thinly. The joint should be as narrow as it is possible to make. The practice of pasting sheets together is not to be recommended, however, for it is difficult to make a really good job. The majority of data can be charted on a single sheet by using a little care in assigning values to the scales.

Using a Pencil in Drawing a Graphic Chart. If a pencil is used first to draw in the curve, mistakes are easily corrected. The final curve is drawn in ink after which the pencil lines may be taken out, together with any other smuts that may be on the graphic chart sheet, with an art gum eraser, which will not injure the ruled lines of the graphic chart sheet, if reasonable care is taken.

CHAPTER XIII

How a Bar Chart is Made

The bar chart is used to give a picture of differences in magnitude. It is perhaps the most easily understood type of chart for its interpretation depends merely upon a comparison of bars of various lengths or heights. The bars may be either vertical or horizontal, but because we are much more accustomed to comparing vertical magnitudes, such as the heights of men, trees, buildings, etc., than horizontal ones, we believe the vertical bars are preferable. On the same chart the bars should all be of equal thickness.

Suppose we wish to show a comparison of the magnitudes of the following figures of a business:

Manufacturing Expense	$567,500
Selling Expense	775,750
Administration Expense	355,500

Fig. 1 shows a bar chart of the figures above. It will be seen that no time element enters into a bar chart of this description. It may simply

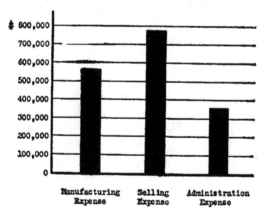

Fig. 1—Bar Chart of Manufacturing, Selling and
Administration Expense for Year Ending
Dec. 31, 1921

be entitled, "Chart of Manufacturing, Selling and Administration Expenses for the year ending December 31, 1921."

A more extensive use of the bar chart is for showing a comparison of magnitudes in which the time element enters, as with the line chart.

The bar chart is arranged fundamentally the same as the line chart. The main difference between the two is that the bar chart represents the

67

numerical values by vertical bars, perpendicular to the "time" base-line.

The steps in making the bar chart are identical with those in making the line chart up to the point where the numerical values of the data are plotted on the graphic chart sheet. These steps are explained in Chapter IV.

We reproduce Fig. 6 of Chapter IV herewith, Fig. 2, which shows the plotted points for the daily receipts of our small business. The next step is to draw perpendicular bars up to the points from the "time" base-line, instead of connecting the points with lines, Fig. 3.

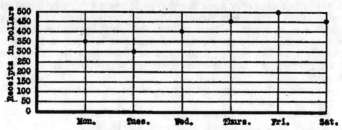

Fig. 2.—Points Indicated on Graphic Chart Sheet

As the purpose of the bar chart is generally to show comparisons in magnitude only, with no regard to trend or rate of change, it naturally follows that it is usually drawn on plain ruling.

Some prefer the bar chart to the line chart because of the fact that the bars, in running up from the zero base-line to their respective points, confine the attention upon the relative heights of the bars, which are the

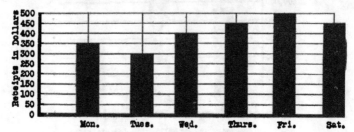

Fig. 3—Completed Bar Chart

measure of the relative magnitudes of the points; whereas if the points are connected with lines, one is very apt to give more consideration to trends, or rates of change as indicated by the slopes of the lines, than to relative magnitudes.

If, however, one realizes that the connecting lines on plain ruling are merely guide lines to assist the eye in following the variations in magnitude from point to point, and that their slopes are measures of increase or decrease in numerical values and not in rate of change, he may use the line chart without fear of being led to false conclusions.

The following is an abstract from an article by John Wenzel,

"Graphic Charts That Mislead," *Scientific American Supplement,* June 16, 1917.

During the last few years there has been a marked tendency towards greater correctness in the form and style of diagrams which have appeared in books, magazines and papers, but many have not conformed to the accepted standards for the making of graphic charts, and are positively misleading.

That the zero line should be shown whenever possible, is a fundamental rule that has been laid down for the making of graphic charts, and yet this rule is frequently broken by those who are engaged in the graphic presentation of facts. Cutting off all the horizontal rulings below the curve line, as is often done, may save space, and the cost of the plate may be reduced if the chart is intended for publication, but by taking away the zero line the correct interpretation of the chart is hindered, and impressions may be created which are misleading. Failure to understand the true significance of the curve line is undoubtedly responsible for the frequent omission of the zero line. If it is desired to show graphically the population of the United States by decades during the last 50 years,

Year	Population
1860	31,443,321
1870	38,558,371
1880	50,155,783
1890	62,947,714
1900	75,994,575
1910	91,972,266

one of the simplest methods is to use vertical bars as in Fig. 4, the heights of the respective columns representing the population at different periods according to the scale at the left. As the eye travels along the tops of these columns, it unconsciously judges their relative heights and constructs an imaginary curve which coincides with the tops of the columns.

Instead of a heavy black column representing the population of each decade, a line could be used, the tops of these lines being joined with a curve, Fig. 5; which by visualizing the population for each respective decade, helps the mind to grasp at a single glance the comparative increase of population.

The next step in the development of the curve line is shown in Fig. 6. Here upon a field with both vertical and horizontal rulings, points marked with an X, representing the population of each successive ten year period, are located on the vertical rulings according to the scale at the left and correspond with the heights of the respective columns and lines in Figs. 4 and 5. By joining these points the curve line becomes the most prominent feature of the diagram.

As the population in 1860 was 31,000,000 and in 1910, 92,000,000 the former was approximately one-third of the latter. Measurement shows the distance from X to the zero line in 1860 to be very nearly one-third of the distance from X to the zero line on the 1910 ruling. The ratio expressed mathematically is approximately 31 is to 92 as 1 is to 3. The

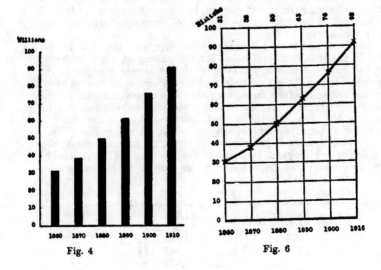

Fig. 4

Fig. 6

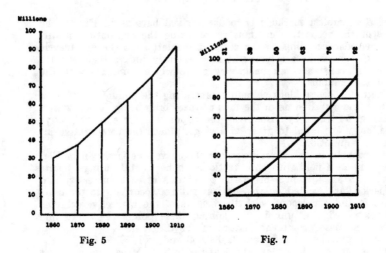

Fig. 5

Fig. 7

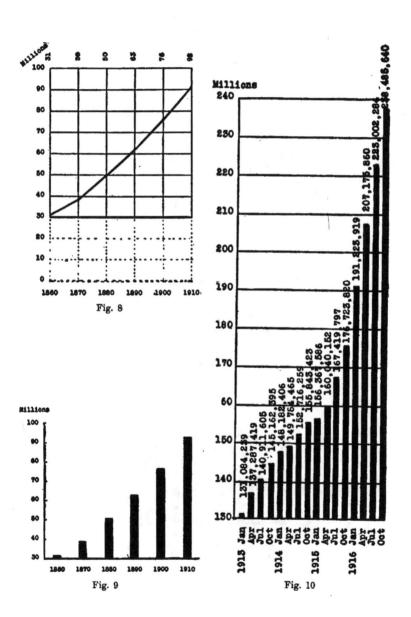

Fig. 8

Fig. 9

Fig. 10

length of the bars in Fig. 4 and the vertical lines in Fig. 5 are relatively correct and convey to the reader a truthful impression of the

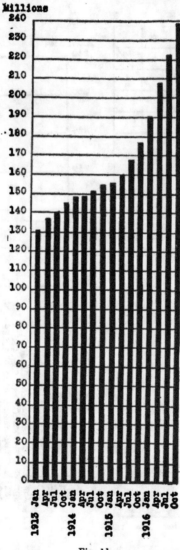

Fig. 11

number of persons who were living in the United States at the times when the census enumerations were made. - Likewise the distance from the zero line to the respective X's gives a correct presentation of the actual population, and thus shows the relative growth from decade to decade. (Relative numerical, not percentage, growth.)

If in Fig. 6 the three lowest horizontal rulings were omitted and the thirty million line were used as the base-line as is shown in Fig. 7, the curve would be identical with that in Fig. 6, but there would be no accurate means of comparing quickly the 1860 population unless an imaginary zero line were constructed as far below the thirty line as the thirty line is below the sixty line, as in Fig. 8. Thirty millions have been cut off of each column thereby practically destroying the measure for comparing accurately the population of one decade with that of another, for few people will take the trouble to construct an imaginary zero line as in Fig. 8.

Fig. 7 has been mutilated and distorted as much as Fig. 4 would be, were 30,000,000 cut from the bottom of each column, making Fig. 4 look like Fig. 9. Returning to the proportion 31 is to 92 as 1 is to 3; if thirty is subtracted from the first two numbers, the result obtained 1 is to 62 as 1 is to 3, is incorrect, the last figure being almost twenty times too small. These figures emphasize the importance of the zero line and give conclusive proof of its necessity for the correct interpretation of a graphic chart.

It seems hardly credible that a financial house which sells industrial securities should wilfully misrepresent the sales of one of the companies whose stock it was trying to sell. Nevertheless, if any one saw the diagram shown in Fig. 10, he would be given an entirely erroneous impression of the difference between sales for January, 1913, as shown in the left-hand column and the sales of October, 1916, as represented by the column on the extreme right.

Only after referring to the scale at the left will it be seen that instead of starting at zero, the scale begins at 130 millions and ends with 240 millions so that the short column at the left which is about one one-hundredth as long as the column at the extreme right, really stands for 131 millions or more than half as much as the column at the right which stands for 238 millions.

Although the volume of sales for each quarter appears in figures at the top of each column the reader may get the real facts only by a most careful examination of the scale and the figures for each quarter. The real function of the graphic chart is to present facts so that the whole story may be told at a glance, and a hasty glance at this diagram would hardly fail to create a wrong impression.

In Fig. 11 the diagram has been redrawn so as to show the correct method of presenting the facts. In this diagram the proper relationship between the sales of January, 1913, and October, 1916, is shown and the impression conveyed is very different from what Fig. 10 would have you believe. In Fig. 11 the scale has been made about half as large so that the two diagrams might be approximately the same size. In this the base line begins at zero instead of at 130 millions. The bar representing the January, 1913, sales is more than half as long as the bar representing the October, 1916, sales which conforms with the actual facts. At first glance the growth as shown in Fig. 11 does not appear to be so great as it appeared to be in Fig. 10, and yet both diagrams present graphically exactly the same data. Fig. 10 shows it in a way which, if not intended to mislead, is likely to do so unless examined with great care. Instead of presenting the real facts so they can be easily and instantly understood, this diagram shows them in a way very likely to be misinterpreted and this is far from the proper function of the graphic chart.

If it will be borne in mind that it is the distance from the top of the bar or from the curve to the zero line that is really significant, then the zero line will be considered as important as the length of the bar or the curve itself, and in drawing a diagram a scale will be chosen that will permit the whole field to appear within the space allowed for the diagram.

Black Tape for the Bar Type of Graphic Chart. Gummed black paper tape for making bar charts may be obtained at most stationers. This tape, about ¼ inch wide, may be cut to the desired length and pasted on the ruled section of the graphic chart sheet. Considerable time is saved by doing this instead of filling in the bars with india ink and a very neat job can be made.

CHAPTER XIV

How a Circular Percentage Chart is Made

The Circular Percentage Chart is based upon the circle. If the circumference of a circle is divided into 100 equal parts, and lines (radii) are drawn from the division points to the center of the circle, the circle will be divided into 100 equal parts or sectors.

The 100 equal parts or sectors represent 100 per cent. Therefore, 20 parts or sectors would represent 20 per cent., 67 parts or sectors, 67 per cent., etc.

Because the circle, divided thus into sectors, resembles a pie which has

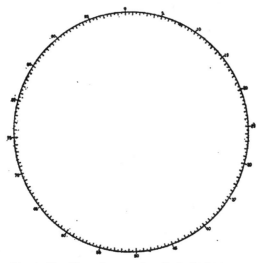

Fig. 1—The Circumference of a Circle Divided into
100 Equal Parts (with subdivisions)

been cut in the usual manner, the Circular Percentage Chart is sometimes referred to as the "pie chart."

The Circular Percentage Chart may be used to show the relation in per cent., of several component parts (the sectors) to their sum or whole, 100 per cent. (the circle). For instance, the cost of manufacturing, cost of selling, and cost of overhead, added together with the profit, are equal to the gross earnings.

For a simple example let us assume the following figures:

YEAR OF 1921

Manufacturing Cost	$502,345
Selling Expense	111,562
Overhead	355,345
Profit	165,080
Gross Earnings	$1,134,332

The gross earnings are 100 per cent. The items that go to make up this total are the following approximate percentages of that total:

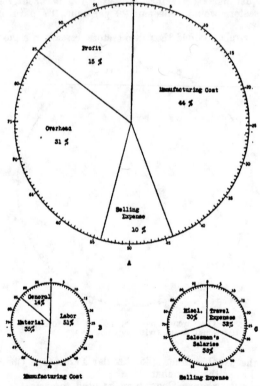

Fig. 2—Circular Percentage Chart

Manufacturing Cost, 44 per cent.; Selling Expense, 10 per cent.; Overhead, 31 per cent.; profit, 15 per cent.; Total, 100 per cent.

To show the above graphically on the Circular Percentage Chart we draw a straight line from the center of the circle to 0 on the circumference. Another straight line from the center to 44 on the circumference will give us a sector representing 44 per cent., or the Manufacturing Cost. The Selling Expense is 10 per cent. We therefore draw a straight line from the center to 54, or the next ten divisions forward clockwise, (44 + 10) and label this second sector, 10 per cent., the Selling Expense. The next is Overhead, 31 per cent. We draw a line from the center to 85, (54 + 31), on the circle which will give us a sector of 31 per cent., the Overhead. The remainder of the circle is from 85 to 100, or 15, which corresponds to the Profit of 15 per cent. This is shown in A, Fig. 2.

Let us assume that the Manufacturing Cost is made up of the following:

Cost of Labor	$256,890	51%
Cost of Material	175,500	35
General	69,955	14
Manufacturing Cost	$502,345	100%

These percentages are shown graphically in B of Fig. 2.

If we also wish to compare graphically the items which go to make up the Selling Expense they may be handled in the same way as the Manufacturing Cost. These are shown graphically in C of the same figure.

Coloring the Sectors. The various sectors of a circular percentage chart may be colored differently to heighten the effect. Either crayons or watercolor paints may be used.

CHAPTER XV

How an Organization Chart is Made

The organization chart is a diagram showing graphically the relation of one official to another, or others, of a company. It is also used to show the relation of one department to another, or others, or of one function of an organization to another, or others.

This chart is valuable in that it enables one to visualize a complete organization, by means of the picture it presents.

There is no accepted form for making organization charts other than putting the principal official, department or function first, or at the head of the sheet, and the others below, in the order of their rank.

The titles of officials and sometimes their names are enclosed in "boxes" or circles. Lines are generally drawn from one "box" or circle to another to show the relation of one official or department to the others.

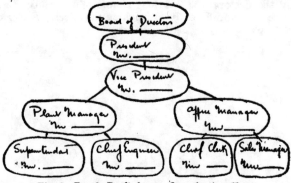

Fig. 1—Rough Draft for an Organization Chart

It is suggested that a rough draft be made of the organization, as shown in Fig. 1. From this the final form of chart may be drawn. Fig. 2 shows a simple type of organization chart constructed on the basis of the rough draft. A chart may be made in as great detail as is necessary, depending upon the size of the company. Copies or blue-prints of the chart may be posted so that all employees may become familiar with the organization.

When an organization chart to show the relation of departments is made, the same general procedure is followed.

It is sometimes desirable to show the activities of various departments by the use of the "flow" chart. This chart will enable one to visualize the steps in the manufacture, assembling and shipping of an article. There is no standard method of constructing this type of chart. A common form is illustrated by Fig. 3.

Organization and flow charts may be drawn on cross-section paper.

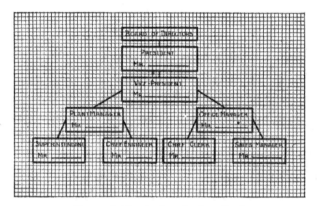

Fig. 2—A Simple Organization Chart

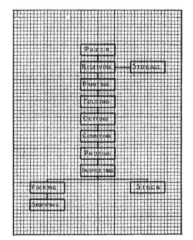

Fig. 3—A Flow Chart

A ruling, 10 divisions to the inch, Fig. 2 and 3, will be found handy in making these charts, as the cross-sectional lines furnish guides of a spacing that is satisfactory both for marking off the "boxes" and for printing in the names of the officers, departments, etc.

CHAPTER XVI

How the Trilinear Chart is Made

The Trilinear Chart is based upon the triangle. While the results that may be obtained from the use of the Trilinear Chart do not limit its form to that of the equilateral triangle, nevertheless because of the simplicity of this type we shall confine our discussion to it.

In any equilateral triangle (one whose three sides are equal), the sum of the perpendicular lines drawn from any point within the triangle to its sides is equal to the altitude of the triangle. For example, in Fig. 1, the sum of the lengths of the perpendiculars from the point P to the sides equals the length of the altitude (the line drawn from the apex, perpendicular to the base-line).

Therefore, if we consider the altitude of the triangle to represent unity, or 100 per cent., the sum of the perpendiculars (measured by the

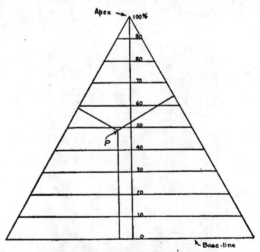

Fig. 1—Altitude of Equilateral Triangle Divided into 10 Equal Parts, Each Part Representing 10 per cent.

same scale) from any point within the triangle to the sides will always be equal to 100 per cent.

This adaptation is made use of in the Trilinear Chart, which is an equilateral triangle with scale lines added. It is utilized to show the relative percentage values of three components whose sum is always equal to 100 per cent. See Fig. 2.

A Use of the Trilinear Chart. The following use for the Trilinear Chart was suggested by Harold Dudley Greeley, C. P. A., New York

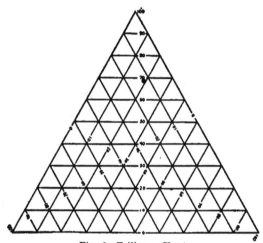

Fig. 2—Trilinear Chart

Manager of the Walton School of Commerce, at a meeting of the National Association of Cost Accountants held in New York City, December 13, 1921.

The Trilinear Chart may be used for showing graphically the relation of three items or percentages, whose sum is equal to a unit or 100 per cent. No matter how the individual values of the three may vary, a single point will indicate these values, provided, of course, the sum is always 100 per cent.

If we desire to show graphically the percentage values of the three items of Material Cost, Labor Cost and Overhead, which go to make up the Manufacturing Cost, or 100 per cent., we may do it as follows: Label the horizontal base-line of the Trilinear Chart, Fig. 3, Material Cost. The side on the left we will call Labor Cost; the one on the right, Overhead.

Let us consider that there is an ideal relation of the three items, as follows: Material Cost, 50 per cent.; Labor Cost, 35 per cent.; and Overhead, 15 per cent. This, we will assume, is the predetermined basis upon which the business of manufacturing may be most economically operated.

This ideal relation may be represented upon the Trilinear Chart, Fig. 3, by a point A, which is plotted, according to the scale shown, at a distance, 50 per cent., up from the Material Cost base-line; at a distance (perpendicular) 35 per cent., from the side marked Labor Cost; and at a distance (perpendicular), 15 per cent., from the side marked Overhead.

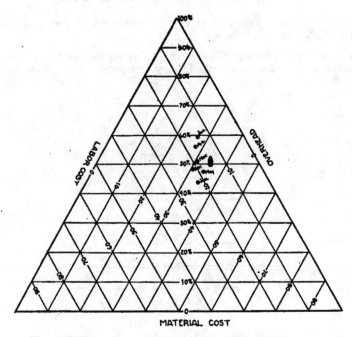

Fig. 3—Trilinear Chart Showing Analysis of Manufacturing Cost

The locating of the point is most easily accomplished by drawing

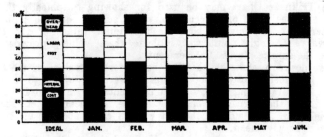

Fig. 4—Bar Chart Showing Analysis of Manufacturing Cost

a line parallel to the Material Cost base-line at a distance of 50 per cent. above it; then by drawing a line parallel to the Labor Cost side at a distance of 35 per cent. from it. Where these two lines intersect each other will be the point A, which, if accurately located, will be at a distance of 15 per cent. from the Overhead side of the chart.

This point A is the bull's eye of the target, so to speak, at which we are to shoot.

Let us assume that the following is the record for a period of six months.

	Ideal Point	Jan.	Feb.	Mar.	Apr.	May	Jun.
Material	50%	42	40	38	37	43	43
Labor	35%	38	36	40	43	39	32
Overhead	15%	20	24	22	20	18	25

These results are plotted and indicated with the names of their respective months in Fig. 3. Thus we are able to see at a glance just how near we approach to our ideal; we note what the trend is from

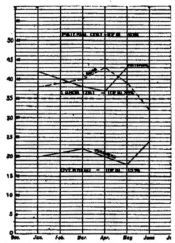

Fig. 5.—Line Chart Showing Analysis of Manufacturing Cost

month to month; and we are furnished with a signal which warns us to investigate if the tendency is in the wrong direction.

Another method of showing the same thing is by means of a bar chart, Fig. 4. The whole bar represents the Total Manufacturing Cost, or 100 per cent. It may be divided up, according to scale, into the three components whose sum equals 100 per cent. The ideal is shown first and then the actual results follow, as they occur from month to month.

Fig. 5 shows the same data plotted on a line chart.

Where there are only three components the trilinear chart has a certain advantage in that single points will show graphically the same facts that require a series of bars on the bar chart and a series of lines on the line chart.

The bar chart and line chart, on the other hand, are not limited to three components, but may be used to show a large number, the only limit being one of clearness, that is, not to complicate the chart so that it is difficult to read it.

CHAPTER XVII

How the Probability Chart is Made

Probability Charts in Business. A great many of the facts with which we have to deal in the every-day world of business occur in groups that follow the so-called "law of probability," or "curve of

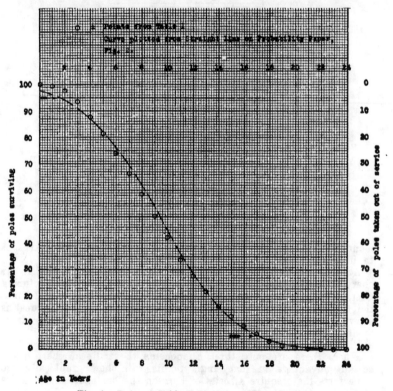

Fig. 1.   Data of Table I Plotted on Plain Ruling.

frequency." This law has attracted the attention of mathematicians to such an extent that within the last hundred years there has grown up about it a very extensive and extremely ponderous literature, and it is only within the last three or four years that the invention of

84

probability paper by Allen Hazen, C. E., has enabled business men to apply the probability law to their problems without mathematical treatment. The purpose of this chapter is to show how to proceed to this end, illustrating the method by an example.

Mr. Edwin Kurtz, in *Administration* for July, 1921, gives a table of the useful lives of some 248,707 wooden telephone poles, the first two columns of which are given below, together with the percentage of poles surviving.

Table I.

Age of Poles Yrs.	Number of Poles Surviving	% of Poles Surviving
0	248,707	100.0
1	247,559	99.5
2	243,336	97.8
3	232,655	93.5
4	218,678	87.9
5	202,045	81.2
6	183,834	73.9
7	164,823	66.3
8	145,563	58.5
9	124,654	50.1
10	104,775	42.1
11	84,011	33.8
12	68,557	27.6
13	54,320	21.8
14	40,541	16.3
15	30,777	12.4
16	22,243	8.9
17	14,584	5.9
18	7,666	3.1
19	3,075	1.2
20	1,277	0.5
21	462	0.2
22	149	0.06
23	47	0.02
24	0	0

Columns 1 and 3 of this table, plotted on ordinary cross-section paper, give the points indicated by the small circles on Fig. 1. A variable curve may be drawn to express the trend of these points, and vertical lines to this curve will give the percentage of poles which have been taken out of service for any particular age. This curve would be the "experience curve" of mortality for these poles, and in form would be somewhat similar to the "normal frequency curve" or "bell-shape curve," as it is sometimes called, of the mathematicians, which expresses the rule of happening, of occurrences subject to the laws of chance. The true frequency curve is symmetrical with respect to its center and a line joining its extremities, and can be plotted by means of the tables given in books on the theory of probability.

The Probability Paper illustrated herewith is so designed that any true frequency curve will plot on this paper as a straight line. The points of the table which were shown on Fig. 1 have been plotted on Fig. 2, to the probability scale, which shows, for instance, for point P, that at the age of 10 years some 58% of poles had been removed, leav-

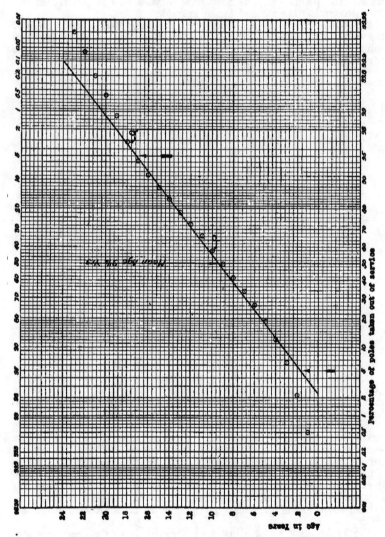

Fig. 2. Data of Table I Plotted on Probability Paper.

ing 42% of survivals. Likewise, for point Q, at 18 years of age about 97% had been removed, leaving 3% of survivals. It should be noted that at both ends of the "life curve" of these poles the points show a marked excentricity, indicating extraordinary conditions. At all events, for the early years, the removal of poles before they are two years old is due to causes different from those affecting the great majority of their fellows. We may consider, for practical purposes, that the data affecting the outside 5% of the survivals at each extremity, or 10% of the whole number of poles considered, may be disregarded as being subject to extraordinary conditions outside of the normal law of frequency. Between these 5% points, marked by vertical arrows pointing upward on Fig. 2, we can draw a straight line which comes remarkably close to all the points within these limits. This straight line on Fig. 2 plots out as the dotted line on Fig. 1. By producing the straight line, Fig. 2, beyond the 5% points, we can get a means of computing the most probable renewals necessary to maintain the lines, under conditions similar to those under which the data were obtained, and disregarding the extraordinary conditions that affect 10% of the existing poles, as explained above.

The fact that the "probability curve" plots as a straight line on Probability Paper makes it possible to construct the curve for any particular problem with a much smaller number of observations and also permits the extension of the curve beyond the limits of the observations with the likelihood that such an extension will be in accord with the facts.

Athough the theory governing the probability curve has been well understood for a long time its application has been limited almost entirely to technical problems owing to the difficulty of plotting it. The designing of Probability Paper has opened up, through its simplicity, a much wider field, and unquestionably there are many problems which may be solved by it that, as yet, have not been worked out. The law of probability is applicable to a wide variety of fields in experimental and statistical work, testing, sampling, insurance, in fact, to all sorts of observations which are not susceptible to exact mathematical analysis but may be solved approximately through their dependence upon the laws of chance.

The following will suggest the nature of the problems in the solution of which Probability Paper will be of great value. Of course it is necessary to have the required observations, similar to those listed in Table I, but in many instances these may be obtained without much effort or expense from data already at hand, but even though special observations have to be made the results usually warrant the effort.

1. Probable heights of workmen, scholars, etc.
2. Probable weights of workmen, scholars, etc.
3. Probable age of workmen, scholars, etc.
4. Probable rainfall, humidity, temperature, etc.
5. Probable death rate in a city, industry, etc.
6. Probable bacteria count in milk, ice-cream, etc.
7. Probable occurrence of new cases in an epidemic, etc.
8. In testing, or inspecting materials—ties, light bulbs, ore, etc., to determine the probable number of imperfect units.

9. In boring for oil, etc., to determine the probability of a successful strike.
10. As an aid in determining the proper amount of reserve—food, money, stock, etc., to meet probable conditions.
11. In the selection of employees to fill certain specifications; to determine the probable total number necessary to be examined to produce the desired selected number.
12. In the classification of employees according to certain standards; probability of how many out of a given total will fall in each class.
13. Bad accounts; probability of what the percentage will be for various sizes of orders.
14. Sales; probability of acceptance of quotations, depending upon amount involved in inquiry.
15. Probable attendance at fairs, expositions, etc., depending upon number and class of people involved.
16. Insurance of all kinds.
17. Apportioning cars, trains, etc., to meet probable traffic.
18. Time study work; probability that a few observations apply in all instances.
19. Department stores; probability of attendance at bargain sales; rainy day crowds, etc.
20. Probable variation in maturity of crops, chickens, cattle, etc.

CHAPTER XVIII

Kinds of Graphic Chart Sheets that May be Purchased [1]

The selection of a graphic chart sheet having the proper ruling and arrangement is important. If a graphic chart is a picture of some event in a business, it follows that this picture should be as clear as it is possible to make it. For this reason, the graphic chart sheet should be developed to standard sizes and standard arrangement. The most popular, and by far the most convenient, is the standard

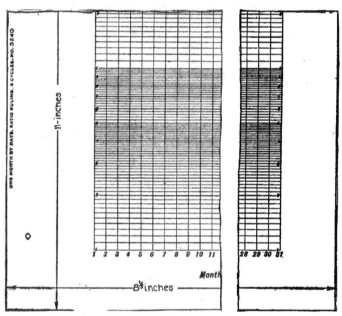

Fig. 2—One Month by Days—Ratio Ruling. 3 Cycles

letter size sheet, 8½ by 11 inches, or multiples thereof.

No longer is the only available graphic chart for business data the "20 divisions to the inch" or "12 divisions to the inch," etc. It is now possible to obtain sheets ruled for "One month by days" or "One year by weeks" or "One year by days," etc. True, the first mentioned graphic chart sheets may be used for many ordinary

[1] Many of the graphic chart sheets described and illustrated in this chapter are manufactured by the Codex Book Company, Inc., 119 Broad Street, New York. They are always glad to send upon request, a complete list of their rulings and to offer suggestions as to which ones are the best to use for specific purposes.

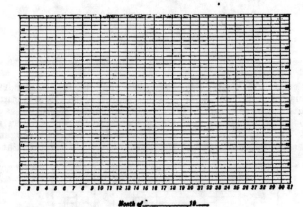

Month of ....19......

Fig. 1—One Month by Days—Plain Ruling
Size of Sheet 8½ inches by 11 inches
Size of Ruled Section 6 inches by 9 inches

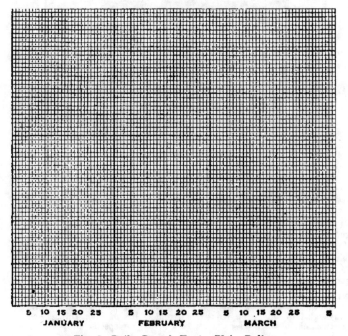

| 5 10 15 20 25 | 5 10 15 20 25 | 5 10 15 20 25 | 5 |
| JANUARY | FEBRUARY | MARCH | |

Fig. 3—Daily Record Sheet. Plain Ruling

charting problems but they do not allow as clear and intelligible a picture to be made, nor are they as easily plotted upon as the others.

Records by Days for Various Periods. The graphic chart sheet, illustrated by Fig. 1 is one of the recent developments in designing a sheet for plotting business data and statistics. It is ruled for one month to be recorded by days. This figure shows the sheet with plain ruling. Fig. 2 shows the One Month by Days sheet with ratio ruling.

Fig. 3 shows the Daily Record Sheet. This is for a period of one calendar year by days. Fig. 5 shows this sheet with ratio ruling. For daily records over a period of any fiscal year, the sheets shown in Fig. 4 and 6 are used. These sheets are 11 by 17 inches and fold once to 8½ by 11 inches.

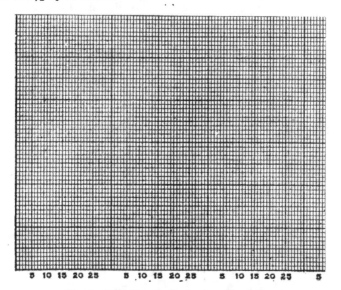

Fig. 4—Daily Record Sheet—Any Fiscal Year
Plain Ruling

For Records by Weeks. Fig. 7 shows a graphic chart sheet for a period of one year by weeks. There are 52 divisions horizontally. Fig. 8 shows a portion of a similar sheet with ratio ruling. The full-sized sheet has three cycles.

Records by Months for Various Periods. For making a graphic chart by months for one year, the sheet shown in Fig. 9 is used. This sheet, "One Year by Months" may also be had in ratio ruling, Fig. 10.

Fig. 11 is for a period of five years by months. This also comes with three cycles of ratio ruling.

Sheets similar to Fig. 11, 11 by 17 inches, covering a period of 10 years, are to be had with either plain or ratio ruling.

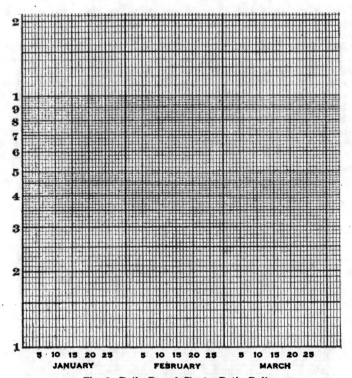

Fig. 5—Daily Record Sheet. Ratio Ruling

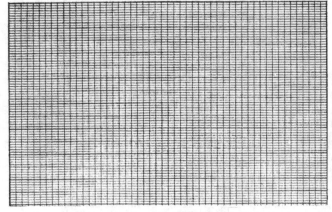

Fig. 7—One Year by Weeks. Plain Ruling

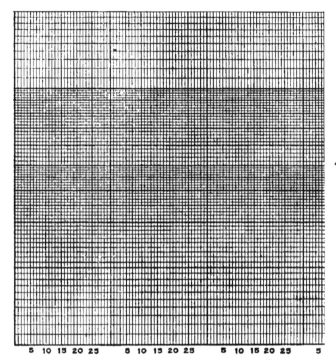

5   10   15   20   25   5   10   15   20   25   5   10   15   20   25   5

Fig. 6—Daily Record Sheet—Any Fiscal Year
Ratio Ruling

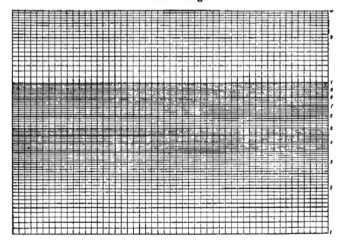

Fig 8—One Year by Weeks. Ratio Ruling, 3 Cycles

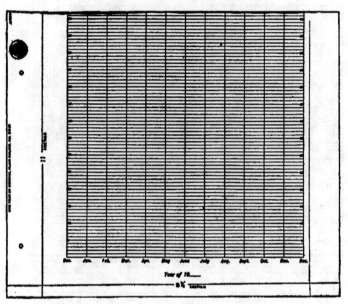

Fig. 9—One Year by Months, Plain Ruling

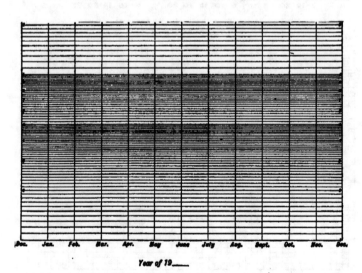

Fig. 10—One Year by Months. Ratio Ruling. Two or Three Cycles

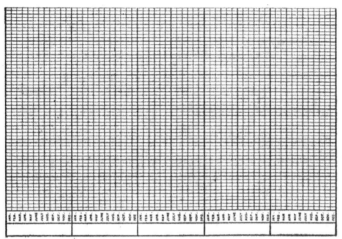

Fig. 11—5 Years by Months

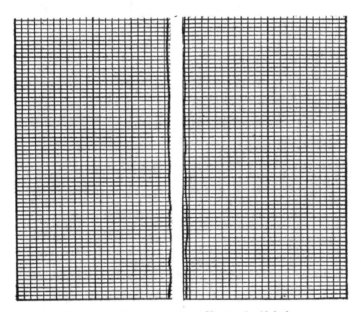

Fig. 12—10 Years by Months—Size 8½ by 11 inches.
14 years, 11 by 17 inches

Fig. 12 illustrates the ruling of a sheet that can be used for records of 10 years by months in the 8½ by 11 inch size, and for 14 years in the 11 by 17 inch size. This does not have the names of the months printed on the sheet, as in Fig. 11.

Records by Years. To plot a graphic chart for a period of years, for instance the index prices of wholesale commodities over a period of fifty years, the sheets shown in Figs. 7 and 8 are used.

For twenty-five years every other line could be used; for ten years every fifth line; for five years every tenth line; etc.

Special Forms. Fig. 13 illustrates a type of graphic chart sheet re-

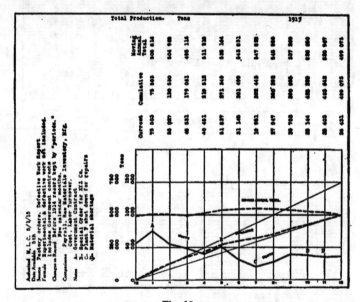

Fig. 13

commended by Arthur R. Burnet in Management Engineering, August, 1921. Mr. Burnet describes it as follows:

"Let us suppose that the subject matter is the monthly production in tons of a certain company for the year 1915. The chart is normally studied while resting on its long side, since that brings the curves into proper position. A person will readily accustom himself to reading all of the typewritten matter without turning the chart around. This method of studying the chart is particularly desirable when charts are examined in the files without removing them.

At the bottom of the chart the numbers 1, 2, 3, 4, etc., indicate the months of the year. The first 12 in the series stands for December of the previous year. The data for this month are plotted on the current chart not only to locate the position of the curves at the end of the previous year but also for the purpose of forming an almost continuous curve extending over 24 months, when two charts are placed side by side and overlapped.

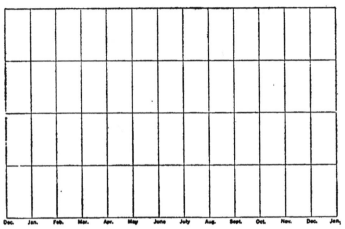

Fig. 14—Wide Margin Chart—Yearly Record by. Months

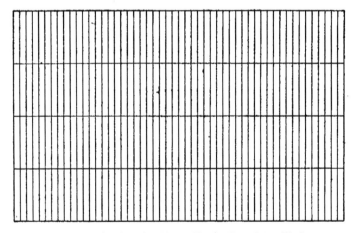

Fig. 15.—Wide Margin Chart—Yearly Record by Weeks

The vertical lines represent the months of the year. Opposite each monthly line at the top of the chart are typed the actual accounting data for the individual months. The executive can therefore use the charts as purely accounting records, or he may visualize the amounts by means of the curves. It is a good rule that the data should always accompany the curves on the same sheet."

Fig. 14 shows the ruling for this type of graphic chart sheet for a yearly record by months. The size of the ruled section is 4 by 6½ inches and it is printed on a sheet 8½ by 11 inches, in the same manner as shown by Fig. 13, so that there is a margin of 4¼ inches at the top and left.

For a yearly record by weeks, the ruling shown in Fig. 15 is used. The arrangement is the same as in Fig. 13.

CHAPTER XIX

GENERAL HINTS AND CAUTIONS

As yet there do not seem to be any generally accepted standards for graphic charts, either as to size or arrangement, these things being governed at present largely by individual preference or habit. A step in the right direction was taken when the joint Committee on Standards for Graphic Presentation published its preliminary report, quoted hereafter. Unfortunately the report comes to the attention of a limited number of people, and those mostly engineers, whereas the use of charts is becoming universal in every business and profession.

The thing which is of prime importance is to have the chart correctly picture the facts which it is intended to show. This depends almost entirely upon the selection of the proper ruling, and it is hoped that no one has reached this point without a clear understanding of this phase of the subject.

There are other considerations, of course, and it is suggested that the following rules of the above-mentioned committee be read over carefully:

1. The general arrangement of a diagram should proceed from left to right.

2. Where possible represent quantities by linear magnitude as areas or volumes are more likely to be misinterpreted.

3. For a curve the vertical scale, whenever practicable, should be so selected that the zero line will appear in the diagram.

4. If the zero line of the vertical scale will not normally appear in the curve diagram, the zero line should be shown by the use of a horizontal break in the diagram.

5. The zero lines of the scales for a curve should be sharply distinguished from the other co-ordinate lines.

6. For curves having a scale representing percentages, it is usually desirable to emphasize in some distinctive way the 100% line or other line used as a basis of comparison.

7. When the scale of the diagram refers to dates, and the period represented is not a complete unit, it is better not to emphasize the first and last ordinates, since such a diagram does not represent the beginning and end of time.

8. When curves are drawn on logarithmic co-ordinates, the limiting lines of the diagram should each be at some power of 10 on the logarithmic scale.

9. It is advisable not to show any more co-ordinate lines than necessary to guide the eye in reading the diagram.

10. The curve lines of a diagram should be sharply distinguished from the ruling.

11. In curves representing a series of observations, it is advisable

whenever possible, to indicate clearly on the diagram all the points representing the separate observations.

12. The horizontal scale for curves should usually read from left to right and the vertical scale from bottom to top.

13. Figures for the scale of a diagram should be placed at the left and at the bottom or along the respective axes.

14. It is often desirable to include in the diagram the numerical data or formulae represented.

15. If numerical data are not included in the diagram it is desirable to give the data in tabular form accompanying the diagram.

16. All lettering and all figures in a diagram should be placed so as to be easily read from the base as the bottom, or from the right hand edge of the diagram as the bottom.

17. The title of a diagram should be made as clear and complete as possible. Sub-titles or descriptions should be added if necessary to insure clearness.

Optical Illusions. For those about to undertake the use of charts the one thing to be avoided is the selection of any method which gives a

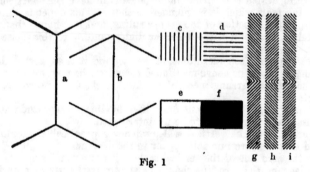

Fig. 1

false picture of the facts presented. There are certain methods of shading, cross-hatching, etc., which should not be used as they are apt to cause optical illusions which are misleading. For example, in Fig. 1 the line a looks longer than b, whereas they are equal. The shaded area c appears to be wider than d, and they are the same. The white area e looks longer than the black one f, but they are equal. Of the bars g, h, and i, g and h seem to be nearest together at their ends; h and i at their middles; whereas all are straight and parallel.

Another type of chart which it is well to avoid is that portraying relative quantities by pictures, as, for example, where the sizes of the armies of different nations are compared by pictures of soldiers. One man, representing an army three times the size of another is made three times as tall as the man representing the smaller army. But as the man grows tall, he grows broad as well, and the eye, in comparing one with the other, includes breadth with height—in other words area—and the tall man looks and is, considerably more than three times the size of the other. There are other similar fallacies and it is well for the person undertaking the use of graphic methods to train himself at the outset to employ only those charts which give a correct analysis of the facts.

The hints which follow are the results of the authors' experience and experiments and are presented in the hope that they may save time and trouble for those about to enter upon the use of graphic charts.

Equipment for Making Graphic Charts. Graphic charts may be drawn with no other equipment than a pencil or pen and a ruler. However, a small drawing board will be found useful, especially when the lettering is to be done by hand. Accurate guide lines can then be drawn for this work. Fig. 2 shows a drawing board, T-square and triangle for making

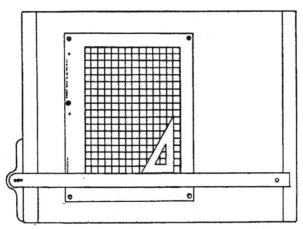

Fig. 2—Drawing Board, T-Square and Triangle

Fig. 3—Newton-Stoakes Pen

a graphic chart. The T-square slides up and down, the triangle to right and left. India ink and a drawing pen will give uniform, permanent lines.

Where it is desired to make an unusually heavy line, the Newton-Stoakes pen, Fig. 3, used with India ink, will save a great deal of time. At one stroke of this pen broad lines can be drawn, which, with the ordinary drawing pen, would have to be outlined and then filled in.

The Title on the Graphic Chart. When a chart is being drawn up the one who makes it usually knows where his data came from, what he proposes to show, etc., but when another person picks up that same chart, or when he himself picks it up again six months later, it will be meaningless unless there is a title upon it to identify it.

The time to put on the title is when the chart is first made and then it will not be forgotten. It should be brief, but should contain all of the information necessary to tell just what the chart is.

It is also advisable to adopt a standard arrangement for the titles, both with respect to their position upon the chart sheets and as to the information they embody. This will save a great deal of time in classifying and filing the charts.

Among the essential facts to be recorded are: 1—What the chart is; 2—Upon what date and by whom it was authorized; 3—The source of the data; 4—The department whose data are charted; 5—References to other charts or files; 6—Upon what date and by whom it was drawn; 7—File number.

Any necessary notes should be included explaining how the data were computed or explaining the causes of any unusual conditions made evident by the curves, or, in general, which will help to give a clear and complete understanding of the chart.

The Most Satisfactory Size of Chart Sheet. There should be adopted a set of standards to govern the sizes of the sheets used in making graphic charts. At present this seems to be determined by individual preference or habit. Some consider the small pocket charts the handiest and others employ great wall charts whose area covers the entire side of a room. For field work and quick reference, especially where it is only necessary to show general trends, the small charts, held together in a ring binder which slips easily into the pocket, are very useful. For general use, however, if the sheets are too small the scale is limited and the range for the time element is restricted. Usually there is not sufficient space to include a suitable title and descriptive notes on the sheet.

Except for purposes of demonstration, as in the class room, lecture hall, court room, etc., the large wall chart has no particular advantages. Everything which can be shown upon it can be shown sufficiently well upon a smaller sheet. The large chart is bulky, difficult to fold, hard to handle and wastes filing space.

It has been proved by experience that the most satisfactory sheet for general charting practice is one 8½×11 inches in size. For special cases a sheet 11×7 inches may be used. This folds once to 8½'×11 inches. It is then possible to file the charts in folders, loose-leaf books, cabinets, etc., with standard correspondence or to include them in reports with typed sheets, making a uniform arrangement of the whole. The 8½×11 inch chart sheet affords a satisfactory scale range and permits the cross-sectional lines to be far enough apart to make the plotting of the curves and the reading of the values easy. Furthermore they are very much cheaper in first cost and require less time to make. For filing and indexing they have the flexibility of a card system with the advantage of allowing sufficient space to satisfactorily present the facts. From our own experience, and from that of others who use them, we believe that the adoption of sheets 8½×11 inches in size, and multiples thereof, as standards will prove most satisfactory to those contemplating the inauguration of a statistical or charting department.

Graphic Chart Sheets in Pad Form. When put up in pad form, graphic chart sheets may be easily handled. The sheets are kept, one kind from another, clean and unwrinkled. The sheets may be bound with thin cheesecloth on the top edge and thin paste on the right edge, 100 to a pad. This method of padding prevents the corners from being curled up by the arm or sleeve when the sheets are drawn upon before they are removed from the pad.

Guide Markers For Punching. The Codex graphic chart sheets are marked on their binding edge with three small circles to serve as guides for punching, Fig. 4. The markers are spaced so the sheet will fit the standard three-ring binder. These markers, of course, do not interfere if another kind of binder is used.

Binding the Graphic Chart Sheets. Among the various methods of binding graphic chart sheets the standard three-ring binder has been found by the authors to be most handy. Fig. 5 shows a completed graphic chart in one of these binders. They may be had with either stiff canvas or flexible leather covers.

Indexes to fit the binders make it easy to file and find the charts.

Reinforcing Graphic Chart Sheets. When graphic chart sheets are punched and filed in loose-leaf binders, they may be reinforced by pasting small gummed cloth patches, one over each hole of the

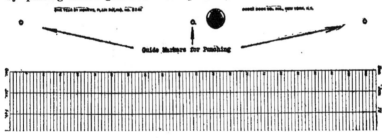

Fig. 4—Guide Markers for Punching

sheet. This is especially desirable if the sheets are subjected to any great amount of handling. The gummed patches, circular in shape, may be purchased at most stationers.

Weight of Paper to Use. It is now possible to obtain most of the rulings described in this book engraved on either thin or heavy paper. Some of the reasons why the thin paper is preferable are: (1) Blueprints may be made. (2) Curves may be traced through from one sheet to another. (3) Transparency permits the comparing of curves on one sheet with those on another. (4) Less bulky. (5) Usually cheaper. On the other hand the heavy ledger paper—(1) Is stronger and more durable, (2) Does not crinkle as easily. (3) Is apt to take ink better as surface is less highly finished.

The Color of the Ruled Section of the Graphic Chart Sheet. Experiment has proved olive-green ruling the most satisfactory. It is easier upon the eyes, especially in artificial light, than some of the more brilliant yellows, oranges or reds. The latter are supposed to make better blue-prints theoretically, but the olive-green has proved entirely satisfactory in practice. The green ruling will also photograph clearly. If it is desired to photograph a chart so that the green lines do not show, this may be accomplished by using a green screen in the camera.

Accuracy of Ruling on Graphic Chart Sheets is of importance if the best results are to be obtained. The graphic chart sheet should be made from a plate that has been engine-ruled. This applies especially to the ratio rulings, where accuracy is essential.

Lettering the Graphic Chart Sheet With the Typewriter. The typewriter provides a quick method of lettering the graphic chart sheet and also assures uniformity. If blue-prints are to be made the graphic chart sheet should be backed with a sheet of ordinary carbon paper. This will increase the clearness and legibility of the figures and lettering on the print.

Making Copies of the Graphic Chart. It is frequently necessary to make copies of a graphic chart, either for distribution among different departments or heads, or to be included with copies of reports, etc. It is also common practice in many companies for the originals to be kept

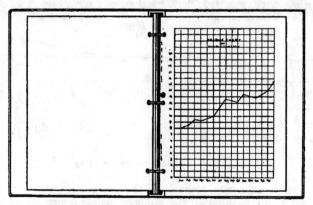

Fig. 5—Graphic Chart Sheet in 3-Ring Binder

on file by the statistical department and never allowed outside their office, which, of course, necessitates copies.

The most frequently used reproductions are the following: (1) Blue Prints and Brown Prints. These are contact prints obtainable only on sensitized paper or cloth. They show white lines on a blue or brown background respectively. Blue Line Prints and Brown Line Prints show blue or brown lines on a white background. They can all be made from originals on translucent material and only to the same scale as the original. (2) Autoprints and Photostats. These are photographic reproductions obtainable only on sensitized paper. Autoprints have black lines on a white background. With photostats the first copy is a negative, white lines on a black background; additional copies either positive or negative. Both can be made from any original which can be photographed and either to the same scale, or enlarged, or reduced. (3) Lithoprints, which are ink prints made from a plate coated with a special gelatin composition. The process is a simplified form of lithography. The reproductions can be printed on any flexible material from any original on translucent material. They can be made only to the same scale as the original. (4) Lithographs, which are ink prints made from a stone or zinc plate. The reproductions can be printed on almost any flexible material from any original which can be photographed. They can be made to the same scale, or enlarged, or reduced. (5) Photo Engravings. These are ink prints made from a line-cut (metal plate). Almost any flexible material can be used to print on and the plate can be made from any original which can be photographed, to the same scale, or enlarged, or reduced.

For full-sized reproductions blue-prints are usually cheapest up to about 125 copies when lithographs become cheaper. Autoprints are cheaper than lithographs up to about 10 copies but are considerably more expensive than blue prints. Lithoprints are cheaper than lithographs up to about 80 or 90 copies, when the latter become cheaper. Where a large number of copies is desired or where it is necessary to keep an original plate for repeat orders, sheets printed from the line-cut are cheapest and best.

For the processes requiring a translucent paper, onion skin, either glazed or unglazed, is generally used but it is very thin and the authors prefer a high grade, light weight bond, as it reproduces perfectly and has the advantage of standing erasure better and does not crack with age.

CHAPTER XX

ESSENTIAL FACTS TO INVESTIGATE AND
HOW TO PRESENT THEM

In many large companies to-day the presentation of statistics has assumed an importance sufficient to warrant the maintenance of a large staff with specially trained supervisors. It is the duty of this department to furnish the managing officials with all of the important operating and financial facts of the business. The whole department is thoroughly organized with certain specialists to gather the data and others to present them.

In contrast to the above are the instances, generally confined to the smaller concerns, where a bookkeeper or clerk is told to "go ahead and get up some figures for the boss." Between these two extremes are to be found all of the various degrees of efficiency in the gathering and presenting of business statistics.

It is generally agreed by experienced statisticians that the graphic chart affords one of the best means for presenting business facts clearly, concisely and quickly. The chart, however, merely furnishes a picture of the data plotted upon it and while it may perform its functions perfectly in the respects just mentioned, it fails in its purpose if the facts it presents are not really essential. In other words, besides a knowledge of how to draw up a chart correctly it is necessary to know what facts will be of value in the economic management of the business.

In every business there are, of course, certain facts essential to its efficient management which pertain to it alone; but, on the other hand, there are many operating and financial facts which it would be worth while for any concern to have on file in easily accessible form. It must be left to the discretion of the manager to decide how far he wishes to go, but we feel confident that what follows in this chapter will help to effect a realization of how important a part statistics should play in economic management and to suggest the general method of approach as well as to outline the details of procedure.

In the May, 1921, issue of *Administration* was an excellent article, "Statistics in Business," by Walter B. Cokell. From this we quote the following:

Statistics are good or bad according to the amount of care given in their collection, correlation, and presentation.

The modern executive spends most of his working hours in an office unacquainted by lack of first-hand knowledge and personal inspection with the results of the activities of the business, even though it may be a compact unit in one building. But most corporations, even of moderate size, maintain their various units in different localities and often operate in several states. It is therefore obvious that if effective control is to be secured, accurate information of the operations of every department, no matter how remote from headquarters, must be promptly furnished to the management in orderly manner so that conclusions can be formed quickly.

Perhaps one of the most recent developments of corporate activity in which statistics play a major rôle is in the building up of research departments. Many banks and bondhouses desire to make intensive studies, both from an economic as well as a financial standpoint, of the particular corporation whose securities they propose to underwrite. Their motive is to determine not only the soundness of the corporation's financial policies but also the value to the public of that corporation's product, the opinion of consumers as to quality and service, and the degree of efficiency at which the factories of the corporation are operated. Likewise, many advertising agencies have started statistical research departments to study the market possibilities of their prospective clients. These organizations are far-seeing and believe the foundation of their own success lies in advertising successful and worthwhile enterprises only. They therefore study a corporation to determine the stability of its management and the quality of its output. Similarly auto truck and tire manufacturers and makers of special engineering equipment employ research experts to study the problems of prospective customers. These specialists then prescribe the particular type or style of their various products best adapted to the existing operating conditions. Here again statistics are used.

In the preparation of statistics five cardinal points should be kept in mind. These are:

1. What statistics to get.
2. The method of securing and the sources of statistics.
3. To whom the statistics are to be presented.
4. The form in which the statistics should be assembled.
5. Getting action on them.

This will be explained briefly as follows:

The statistician's job in a business organization is to determine what data are necessary for the use of executives, to anticipate as much as possible their requests for information and, on the other hand, to avoid collecting data which will be of little or no use. For example, some large companies, might wish for detailed weekly reports of the sales of each article in each county of the field, while for a smaller concern a monthly report of the sales of the more important articles by states would be sufficient.

Oftentimes statistics of certain kinds may be very useful for a time and later of less use because of changed conditions. An executive may wish for weekly labor reports during a period of reorganization in order to study the change in the size and character of the personnel. Later on, however, after the organization is well established a report of this kind rendered monthly or even quarterly may be sufficient.

In collecting his data a statistician must be careful of his sources of information. When possible and when the cost is not prohibitive, data should be taken from original sources, unless the statistician has confidence in a report already issued and knows how it was compiled.

Care must be taken to determine the accuracy and the exact meaning of the data at hand.

The statistician must remember for whom he is preparing the information or the particular use to be made of it. A production manager is interested primarily in the costs of production. Only to the extent that they would affect his costs or necessitate changes in processes or the like would he be interested in sales statistics. Hence, only in exceptional cases would he need sales figures. On the other hand, a sales executive wants to know all about his sales, the proportion coming from each salesman or territory, the ratio of selling expenses to sales, and the ratio of sales to expected quotas. But he would not care to know so much about the detailed cost of the product except where it might affect selling prices.

This depends largely on the purpose the figures are to serve. In the interests of economy and as an aid to making proper comparisons, a statistician should endeavor to standardize his reports. Always, however, special statements prepared to set forth certain facts will require a form of report to suit the figures. Data should be assembled into the proper units. These units can then be combined in any manner required to give the proper totals. To illustrate, the sales manager may want to know the total sales of a certain article and the sales of this article by states, zones, or salesmen. If an analysis of the sales is first made merely to get the grand total, the work will have to be done over to answer the second request. On the other hand, if the data are first compiled of sales by salesmen, these figures can easily be built up into sales by zones, states, or other geographic divisions as required, without making a second analysis of the basic data.

If the first four points have been carefully worked out this last step will be easy. For, if the purpose of the figures has been kept in mind, if the correct plans of collecting the figures from the proper sources have been followed, if the needs of the persons for whom the figures are to be presented have been kept in mind, and if the figures have been assembled in a manner that will be clearly understood by the person to whom they are to be presented, then the action necessary is usually so plainly indicated that no argument is needed. The facts will speak for themselves.

The executive of to-day, no matter what line of business he is in or what his particular position is, should have, in the writer's opinion, a general knowledge of the fundamental principles of four subjects in addition to his specialized training. These subjects are economics, the science underlying all business; law; engineering; accounting and statistics. The necessity of a knowledge of the first three is fairly well established. It is only rather recently, however, that business men in general have begun to recognize the value and necessity of accounting records and reports. The science of statistics is still a new subject to all but the most progressive executives.

Graphics can be said to represent the last word in statistics, for usually, after a table of figures has been set up, the next thing is to "illustrate" it by means of a chart.

Dwight T. Farnham, in a series of three articles published in *The Engineering Magazine*, August, September and October, 1916, entitled

"Visualizing the Essential Facts of a Business," "Scientific versus Intuitive Administration," and "How Graphic Control Facilitates the Fixing of Profits," points out what he considers to be the facts which the executive should receive in simple and effective form. From these the following has been quoted:

Cost figures may have historical interest but they are not worth one-tenth the expense of assembling unless they aid actively in the administration of the business. The problem of the executive, then—once his organization is perfected—is to secure live data covering the exact conditions of the business at all times. These data should be arranged so as to give him all the facts, subordinated according to their relative bearing upon net earnings, and do so with the least demand upon his time. Furthermore, these facts must be so exhibited that the general laws underlying the business may be easily and accurately deduced and standards of accomplishment set which will be a continual incentive to greater accomplishment.

If a picture can be arranged which will give him at a glance the exact state of his whole business, with detail subordinated in the order of its importance, so that the large and important things stand boldly in the foreground where they cannot be overlooked, and the less important facts, though present, receive attention only when necessary, the executive is in a position to grasp and to direct his whole business with an intelligence and sureness of touch which insures his stockholders getting the fullest return on their investment.

A few years ago some of our more progressive corporations began showing the fluctuations of their sales in different districts by means of graphs. Later the same method was applied to costs, but it is only recently that anything like a comprehensive scheme covering the whole business has been worked out and put into effect, Within the last two years several large corporations have gone a step further. They have not only shown by means of graphs what they *had done* and were doing, but also what they *proposed to do.* And the most remarkable thing about these prophecies by graphs is the regularity and exactness with which the ideals aimed at have been realized.

It is absolutely necessary, first of all, to analyze existing conditions. This is extremely difficult for the executive to do with any degree of precision where the usual types of balance sheets and cost records are in use. When these are placed on his desk at the end of the month, he glances through them curiously and if, in the light of what he happens to remember about past performances, the showing seems satisfactory, he drops the sheets and returns to his routine. A sudden drop in profits or a rise in costs presages a call on either the sales or the manufacturing department, or both, for an explanation. Usually these departments know what kind of an answer will satisfy the "old man." The ancient bogies of competition or of interrupted production, are paraded forth and the "old man" goes back to his desk. This is all that happens unless a prolonged period of low profits ensues. In this event the directors come in and hold a post mortem. After a critical review of the office rent, the amount spent for postage and telegrams and the salaries of the stenographers (I have seen board meetings over exactly this sort of trash), a period of retrenchment is considered vital. The office force is thereupon cut down and the department heads are deprived of their stenographers. A man here and a man there, on the "non-productive" side of the ledger, is dismissed. When they get through they don't know any more about the real cause of the falling off in business than they did before. Worst of all, the retrenchment has applied the axe, perhaps, to the only part of the organization which could by any chance tell them.

For such summary action, however, the financial managers cannot be wholly censured. The statistics prepared by only too many cost-keeping departments are misleading rather than enlightening to the executive, and it may be the less of them the better. Masses of figures come to him each month which cannot possibly be remembered until the next month. As a result conditions each time have to be considered largely by themselves or as contrasted with some other single month. The very volume of the figures makes it a physical impossibility to lay out on a table and to compare the statistics for more than a few months at a time. It is not in this way that the laws underlying fluctuations in costs, in sales prices, or in profits can be deduced. Yet predictions as to future conditions are an impossibility unless these laws are known.

Conditions are, of course, different in every business, but the underlying principles are identical in all. Were this not so, bankers would not be able day after day, by a few well put questions, to lay hold, as they are accustomed, of the essentials of countless business ventures.

Under the following heads the facts which an executive should have before him at all times in order to administer the business efficiently are given in skeleton form in the order of their greatest importance.

I—*Dividends*—It goes without saying that dividends—large, regular and frequent—are the *raison d'etre* for the existence of any business, since it was the hope of an adequate return upon their money which induced the stockholders in the first place to invest. Dividends are directly dependent upon earnings, so that the first leaf in the "Bible" of the executive should show graphically the fluctuations in earnings over the period of the concern's existence.

II—*Profits*—Since profits consist in the difference between the selling price and cost price, the second section should show by means of curves the average monthly selling price per unit of the major products as compared with their total cost sold. The space between these two curves at all times represents the profit. Low profits may be due to a low selling price or to a high manufacturing cost. By watching the fluctuations in the two curves the executive at once can determine which of the two halves of his organization needs his attention and assistance, and he can throw his strength to the weakest point.

III—*Sales*—Graphs showing the total quantity of sales, the distribution of quantities by districts and subdistricts, show the executive and the sales manager just how well each portion of their organization is doing as compared with past performances and just which portions of the territory need attention. The prices obtained in each section of the country are averaged each month and the data so presented that every effort may be intelligently and consistently made to hold prices to a maximum. In a similar manner sales expense is kept track of—subdivided as common sense suggests.

In concerns which have applied the principles of scientific management to their sales department—rewarding their salesmen and sales-managers exactly in proportion to what they accomplish, quantity sold, price obtained, and expense saved—the "efficiency," or percentage of attainment of the standard set on these points in each territory is platted so that the executive may see at a glance just what is being accomplished in each territory in proportion to what careful analysis of local conditions in each district has determined should be accomplished. This standardization simplifies the work of the executive considerably, since the same results obtained from the application of scientific management to the factory—such as the increase in the employees' interest in their work, the introduction of team play, etc., etc., renders less active supervision upon the part of the chief executive necessary, and the percentage system of recording results boils down all results to a common denominator so that much less of his time is required to grasp the exact state of affairs.

IV—*Manufacturing Costs*—Should on examination the sales graphs show that low profits during a given period were in no way attributable to avoidable faults in the sales department, the executive would naturally next turn his attention to the manufacturing department. It is not our purpose to indicate in detail how a complete cost system may be reproduced graphically. These, however, are the principal questions which the executive probably would be prompted to ask: (a) Was the total cost of manufacture high or low as compared with past costs? (b) As compared with the standard costs? (c) Was the output unusually high or unusually low? (d) For how much of the variation in costs was the output responsible? (e) Was a period of low output followed by merely a rise in the indirect labor, supervision, rent, general expense and various other sorts of overhead or did it extend to the direct labor, showing, that men were kept on unnecessarily in slack times by foremen "just to be a good fellow"? The answer would at once determine the executive's course of action. (f) Was the variation in cost due to labor or material?

With such questions as these answered—in fact with the answers so arranged in advance that they strike the executive forcibly, unavoidably, relentlessly, and regularly every month—the necessary action is taken or the executive cannot avoid the admission, even to himself, that he is incompetent.

V—*Material Costs*—An increase in the cost of raw material often takes the rise in costs outside the executive's control. At the same time he should know exactly to what extent this rise affects his cost of production. Such knowledge either drives him to raise the selling price— and for a legitimate reason—or to attack with renewed vigor some department which seems capable of reducing its operating costs sufficiently to offset the increased material costs. In other words he is driven to make an extraordinary effort to meet the changed condition by a knowledge of its exact seriousness and his stockholders can rest assured that everything possible has been done to safeguard their interests.

VI—*Elimination of Waste*—Every business has its own particular sort of rat holes, through which its profits are carried piecemeal, and in quantities hardly noticeable at the time, but which aggregate thousands every year. The best way to plug these sources of loss is by accumulating data in regard to them and then keeping this data prominently before the executive. Possibly a graph showing the total rejections by customers is required; or a graph showing the number of pieces spoiled in each department. Knowledge of this sort ends very shortly in a raid upon a particular department and a general reorganization along more efficient lines. The result is a sharp drop in the rejection curve. Sometimes it is the production of seconds which needs watching, or perhaps the scrap produced, shown by departments. Frequently a curve showing the number and causes of shutdowns leads immediately to action which saves the company thousands of dollars. The mere reduction of such matters to figures is invariably beneficial. Couple such tabulation with an effective graphical presentation and the executive cannot fail to note the consequences of the losses in dollars and cents. It is his fault, then, if he fails to act at once and strike straight at the root of the trouble.

VII—*Inventories*—Very often, too, the accumulation of manufactured stock becomes a serious menace to a business. The executive may know in a general way that a large quantity of a certain variety of product is on hand; he may know furthermore approximately what the total investment in finished stock aggregates. The important matter, however, may be the size of the stock of each variety as compared with the stock at the same season previous years, or at a period when market conditions or manufacturing conditions were similar. Certain grades of product have a way of accumulating in many plants, especially where the market demand is not as great as the production of these grades. With the facts all before him the executive is in a position to dispose —by bargain sale or special attention—of all grades to good advantage as they are manufactured; while if the facts do not come to his attention until the accumulation of stock is large, a forced sale at a sacrifice may be the only remedy.

VIII—*Labor Turnover*—This is another matter which few executives watch closely. The figures are sometimes available in the office of the employment department or can be assembled from the records of the timekeeper. In most cases, however, executives do not know how many men they are hiring annually to maintain their force. When this insidious source of loss is called to their attention they will usually admit that it costs them $50 to $125 to break in a new man. Especially will they make this admission if they can be induced to recall a few of the mistakes costly to their employers which were made when they were serving their own apprenticeships. The figures are usually appalling when first compiled. It is not at all unusual for a business to have a labor turnover of from 100 per cent. to 600 per cent.! A concern employing a thousand men, whose normal turnover should not exceed 50 per cent., has one of 250 per cent. Assuming that it costs them fifty dollars to break in each new man—and fifty dollars is low if you consider the spoiled material, and the time of foremen and fellow employees taken up in teaching a new hand the ropes—the annual loss amounts to $100,000!

An examination of the labor turnover often brings unexpected inefficiencies to light.   One concern which the writer investigated had a turn-over of over six hundred per cent.—about four hundred per cent. above normal, considering the conditions in the local labor market.   It was thus losing about fifty thousand dollars a year unnecessarily.   The searching investigation which followed disclosed the fact that a superintendent was receiving a rake-off from an enterprising employment agency for every man he hired!   The sudden and enforced exit of the super-intendent resulted in the turnover percentage dropping to normal.   Thou-sands of dollars are being saved every year in consequence.   A graph which keeps the labor turnover before the executive makes such oc-currences as this impossible.

IX—*Safety First*—The "safety-first" movement has ceased to be a humanitarian fad and is now generally recognized as fully justified for sound business reasons.   The cost of damage suits, hospital bills and pensions to say nothing of the cost of "finding a place" for disabled employees during convalescence, breaking in a new employee on short notice, having the crew disorganized by the absence of an employee for a day or two and the general fall in morale in a factory where accidents are frequent, is sufficient to make the reduction of accidents a matter of considerable importance to the executive.   Tabulations showing the causes of all accidents each month and a graph to keep the state of affairs before the executive continually just as inevitably leads to the elimination of the causes of accidents as the elimination of the accidents leads to increased earnings in the business.

X—*Daily Curves*—Finally, every business has certain vital operations —it may be the pounds of coke required to produce a ton of steel or the specific gravity of a chemical—upon which depends to a large extent the success of the business.   Graphs which keep such quantities or such percentages daily before the executive bring his attention and assist-ance at once in case of need.   The correction is immediate and the saving thereby effected may be far-reaching.

Suggestions for Graphic Charts that Aid Business Control.   The following is not intended to be a complete list of charts necessary for executive control.   It is given in the hope that it may suggest ideas and open up the way to a better realization of the possibilities of the graphic method of presenting statistics.

These items may be charted by the time intervals—days, weeks, months, or years—calculated to afford the most efficient control.

I. GENERAL
    A. General Commodity Prices.
    B. Specific Commodity Prices Applying to the Business, such as Iron, Copper, etc.
    C. Labor, Rates of Wages.
    D. Materials, Prices.
    E. Supplies, Prices.
    F. Temperature.
    G. Humidity.

II. FINANCIAL
    A. Accounts Receivable vs. Collections (gives lag).
    B. Accounts Payable.
    C. Cash on hand.

D. Call Money Rates.

E. Bank Rates for Commercial Paper.

III. LABOR

    A. Daily Force on Hand.

    B. Labor Turnover.

    C. Absence from Illness.

    D. Absence from Accidents.

    E. Daily Wage Total.

    F. Results of a Series of Time Studies on Consecutive Operations.

    G. Results before and after Showing Man His Performance.

    H. Fatigue Study.

    I. Bonus Curves.

IV. OUTPUT

    A. Daily Production vs. Manufacturing Schedule.

    B. Productive vs. Idle Time, Men and Machines.

    C. Unit Costs of Production.

    D. Rejections of Product Due to Spoilage.

    E. Number and Causes of Shutdowns.

    F. Progress Charts.

    G. Relation between Raw Stock on Hand, Ordered and used by Factory.

V. MISCELLANEOUS COST RECORDS

    A. Cost of Manufacturing; High or Low Compared with Past Cost.

    B. Cost of Manufacturing; High or Low Compared with Standard Cost.

    C. Raw Material Costs vs. Production Costs.

    D. Cost of Packing and Shipping.

    E. Cost of Lighting, Heating, Transporting, etc.

    F. Daily Overhead Factor.

VI. SALES

    A. Total Sales vs. Sales "Quota" or Estimated Sales.

    B. Sales by Districts, Cities, States, etc.

    C. Comparison of Sales of Different Articles Manufactured.

    D. Subdivision of Sales Expenses.

    E. Performance of Salesmen.

    F. Daily Orders Received, Filled and Unfilled.

VII. ADVERTISING

    A. Inquiries Received vs. Orders Received.

    B. Results from Various Methods of Advertising.

    C. Ratio of Advertising Expense to Sales Resulting Therefrom.

The Following Pages illustrate the use of graphic charts in connection with various departments, advertising, sales, etc. The sections are arranged alphabetically for easy reference. Some of the material presented has been abstracted from current business and industrial magazines, and the charts illustrating these articles are not always plotted on the most suitable ruling. Charts intended to show percentage variations have been plotted on plain ruling. In spite of this fact, however, they were selected because they suggest valuable ideas for the use of charts, and, after all, the ideas are of more importance than their presentation, for by the time the reader reaches this point he will have mastered the function of the various rulings, and, in following out the ideas suggested, will be able to make his own charts correctly.

The charts shown and described hereafter are by no means a complete collection of all the charts that might be gathered in connection with the various departments, nor is it intended that they should be. In fact, we do not feel that there is anything to be gained from an indiscriminate collection of slightly different examples for no two businesses are exactly alike, and, in all probability, no one would find illustrated just the chart he required.

Therefore, it is rather to suggest than to specify that we have shown the following examples. One of the most interesting things about charts is the fact that one is constantly finding new uses for them in connection with his particular problems. They stimulate the imagination. We have selected illustrations which we believe, by their diversity, will make evident the great field of usefulness that exists for this unique method of presenting facts.

CHAPTER XXII

GRAPHIC CHARTS IN ACCOUNTING

The graphic chart may be made a valuable part of the accountant's report whether it be an annual report of the business, a report for a single department, or a report of the analysis of some particular phase of the business. In fact, any financial report, if illustrated by charts, will be greatly enhanced and thus permit a more comprehensive understanding of the conditions reported.

The stockholders of a company are concerned with the dividends on

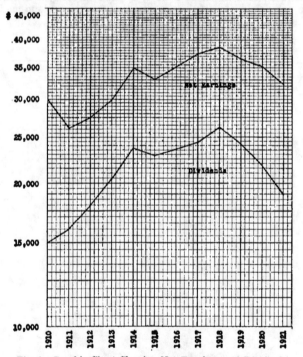

Fig. 1—Graphic Chart Showing Net Earnings and Dividends

their stock. The graphic chart, Fig. 1, shows net earnings and dividends over a period of years. The numerical difference between the curves represents the margin allowed for emergencies. A study of such a chart will show whether the margin is being upheld and the slope of the lines will show if it is being held at the same rate. This chart will also show the rate of increase or decrease in net earnings and the tendency.

Another chart, Fig. 2, gives the graphic record of cash balances at the end of each month for a year. The fluctuations are easily and accurately noted, and the picture presented is infinitely more interesting and clear than the tabular data from which it is plotted. Being on ratio ruling, a correct picture of percentage variations is obtained.

Fig. 3 shows a graphic chart for payroll analysis. By having the three curves on a single sheet a direct comparison may be made. The degree of divergence between the curves, Number of Men and Wages Paid, is a measure of the trend of the average wage—whether it is increasing or decreasing; and that between Number of Men and Hours Worked is a measure of the length of the average working day—whether it tends to increase or decrease. Such curves tell a very interesting story if plotted over a period of ten years or more.

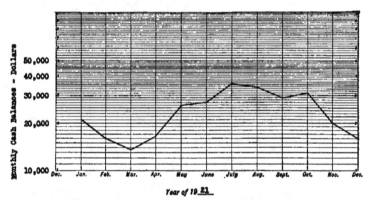

Fig. 2—Graphic Chart Showing Monthly Cash Balances for One Year

The following is taken from an article by W. F. Bloor, entitled "Value of Graphics in an Accounting System," published in *The Journal of Accountancy*, June, 1921.

The underlying reason for the growing demand and value of accountancy is the realization of the business man that an accurate analysis of his business is as essential to progress as a carefully planned sales or production organization and policy. The decisions and judgments of the successful man of business today are not, as in the past, based upon guess work or a "hunch," but upon a thorough knowledge of the past, present and future.

The method of presentation is, therefore, of much importance. Some executives, for example, require a detailed record of operations, while others need only a statement of the outstanding facts. A busy factory or sales manager is not vitally concerned with an increased cost of machine supplies in one department nor with excessive traveling expenses of a salesman. These are facts which concern only executives further down the line.

There are four general methods of presenting accounting data:
(1) By written report.
(2) By tables of figures.
(3) By graphics.
(4) By combinations of (1), (2) and (3).
Presenting business and accounting facts by a written report and statistical tables is often necessary, particularly when much detail and

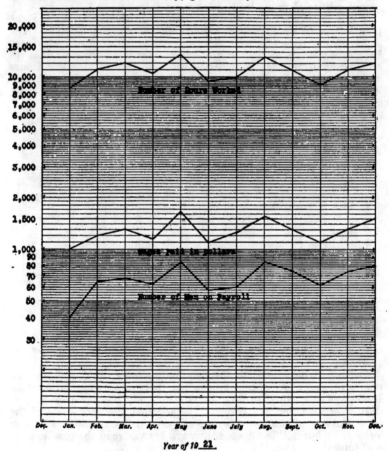

Fig. 3—Graphic Chart for Payroll Analysis

lengthy analyses are desired. However, when trends or tendencies or a continuous story of vital data only is desired, the graphic chart has no substitute.

During a period of business depression, when economy has become the watchword in industry, complete control of the situation has been in the hands of those who are responsible, largely because they have the facts constantly before them. If the waste-cost curve, for example, of a certain production department turns sharply upward one month, the factory manager realizes it to be a danger signal; and if it continues the next month, there is no escaping the conclusion that something must be done to turn the curve downward. A conference with the superintendent, showing him the chart and the disastrous results should the present condition continue, usually brings the desired results.

It appears evident from the foregoing that graphic methods of presenting facts are becoming essential in all business and other activities of man. No matter how small the business unit, some consideration should be given to a method of presenting the facts graphically. In a small organization the system, of course, would be relatively simple as compared with a complex and complete system for a hundred-million-dollar corporation.

Fig. 4—Organization Chart of an Accounting Department

Fig. 4 is an organization chart of an accounting department of a manufacturing concern, which, with slight adjustments, should apply to any business. Conditions are, of course, different in every business, but the underlying principles are identical in all. Under the department manager are three main subdivisions, namely, general accounting, cost accounting and statistics and graphics.

The function of the statistical and graphic subdivision is that of collecting, compiling and analyzing statistics of production, costs, sales, profits, etc., furnished by the general and cost accounting sub-divisions, and presenting the data to the management in the most efficient manner.

External data may be coördinated with the internal statistics through research study, which is the function of the research section. Comparisons of commodity prices with wages, production of raw materials with market prices, sales with bank clearings, etc., are valuable information for executives. In fact, external data should be included as part of every graphic system. Business and economic conditions and movements in general have direct and vital bearing on the future of any Business.

CHAPTER XXII

GRAPHIC CHARTS IN ADVERTISING

The graphic chart may be utilized to perform two distinct functions in connection with advertising. The first is for purposes of efficient management or control; the second is for purposes of demonstration to prospective customers.

Besides the charts showing the expenses, etc., which are equally valuable in any department there are those which apply directly to the activities of the advertising manager. He wants to know, for example, if his publicity methods are paying—how the returns from a full-page "ad" in one magazine compare with the returns from a half-page "ad" in another; whether his maximum returns result from using 400 newspapers, or 300, or 600; whether his net returns warrant the use of preferred locations, color work, etc. All of these things may be determined by collecting the necessary data and showing the results on a chart.

By keeping a chart such as Fig. 1, the manager can tell whether his advertising is bringing in the right sort of inquiries, i. e., the kind that result in sales. This chart shows the total inquiries received over a period of time and the total number of orders. Should the percentage of orders drop with respect to the inquiries, which condition can only be correctly shown on ratio ruling, there is something wrong. It may be due to the fact that the advertising matter is misleading or due to mismanagement of the department that handles the inquiries. At any rate, it starts an investigation to determine the cause.

The total sales and the total advertising cost make an interesting comparison, Fig. 2. If the sales keep up when the advertising expenditure is lessened, the inference is that efficient methods are being used by the advertising manager. If the cost of the advertising shows an increase while the sales do not materially increase, the advertising methods should be investigated. If the advertising is uniformly efficient, the curve of sales will lag behind the curve of expenditures to a certain extent, but will parallel it.

In comparing cost of advertising with sales, only the sales resulting directly from the advertising should be used. The efforts of the salesmen and the sales department cannot, of course, be credited to the advertising unless the two are planned in direct conjunction.

Similar charts may be plotted for the total efforts of the company, or may be made for each territory. In this manner it is possible to compare the efficacy of various advertising methods on various communities.

Graphic Charts in Direct Advertising. Graphic charts are appearing with a rapidly increasing frequency in connection with advertising matter in magazines, form letters, etc. If it is desired to compare the cost or operation of one thing with another there is no method equal to a chart

120

for showing it clearly and concisely. A machine may be pictured by a photograph and its performance may be pictured by a chart. A combination of the two makes a very forceful appeal. The only reason they

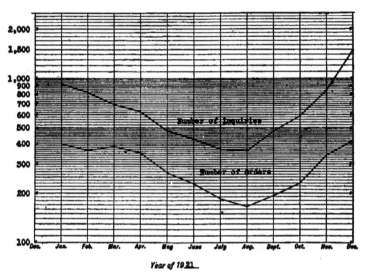

Fig. 1—Graphic Chart Showing Relation Between Inquiries and Orders

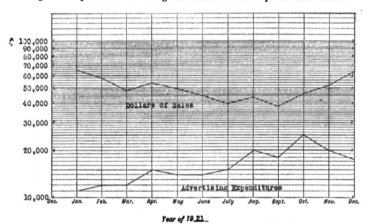

Fig. 2—Graphic Chart of the Relation between Sales and Advertising

have not been used more frequently in the past is because they have not been understood by the layman, but the general public is rapidly coming to appreciate their significance and to take an interest in their message.

The Strathmore Paper Company used a bar chart in their advertisement in System for September, 1920. Fig. 3, a reproduction of the chart, shows the relative magnitudes of the different costs which go to make up the total cost of a letter and emphasizes the fact that the cost of the paper involved is very small.

The following is from an article by A. H. Richardson in the April 1, 1921, issue of Industrial Management.

To grasp accurately the relationships between quantitative values it is usually necessary that figures be presented in graphic form. Advertisers are beginning to realize this and to turn to the use of charts and diagrams

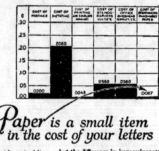

Fig. 3—Bar Chart Used in Ad of Strathmore Paper Co.

as the most effective way to tell the side of their story that has to do with statistics.

Interest in an article advertised can often be aroused through the presentation of background facts which are of general interest to the reader, and at the same time definitely related to the product. Much information is available to draw on and as this is largely statistical there is presented considerable opportunity for the use of charts. A number of examples are given to illustrate the various kinds of backgrounds that were found.

Fig. 4 suggests the possibility of presenting historical data in chart form. There is no technical product or industry that has not a history, and for much of this it is possible to assemble figures and to present them in the form of charts, which will enable the reader to gain a broad setting for the article advertised.

Effective comparisons can often be made between conditions in one industry and business conditions in general. No technical product with respect to its manufacture and use is isolated from general economic

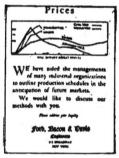

Fig. 4

Fig. 6

Fig. 5

Fig. 7

and social conditions. An example of the use of comparative cost and price figures is shown in Fig. 5 where the changes in the price of Goodyear Tires is contrasted with material and labor costs over the same period.

There is considerable significant material relating to standards of performance which can be shown effectively in graphic form. The best example found is shown in Fig. 6 which pictures the distances in which a car going at different speeds should be able to stop.   Every car owner is interested in being able to stop his car and interest in Thermoid Brake Linings presumably follows.   .

A group of charts which seemed to the writer to be of particular interest were those which dealt with the problems of the prospective buyer.   It is an increasingly common thing to find among the technical advertisements a discussion of the conditions under which a man may be carrying on his work and the advantages to him of doing it in some other and more efficient manner.   Such advertisements are, primarily, educational and are, to be found usually in presenting the merits of a new kind of machine or a specialty.   It may be assumed that the reader is not familiar with the machine or at least is not convinced that it is a thing he should be interested in.   The task is to get him to realize the relationship between the machine in question and his own peculiar problems.   Charts are of great value here because of the chance they offer of picturing a situation which will be recognized at a glance.   Take, for example, the curve in Fig. 7.   Every business man will recognize the peak of figure work that comes at certain times during the week or month.   .

There are some kinds of information which suggest certain rather fixed methods of presentation.   For example, the line chart seems best adapted for showing performance, column charts for showing growth of certain kinds, "pie charts" for showing simple percentage comparisons, etc.   But there are many kinds of information that do not fall into the conventional groups, and the proper selection of some variation of the conventional graphic forms becomes important.   Each chart must be a matter of individual study in order to select the graphic device that will tell the story most effectively.   This presents the opportunity for the display of imagination, though too much imagination is a constant danger to be avoided.   So it is important that accepted underlying conventions be followed.   Graphics used in advertising should be truly graphic.   This means they should be reduced to simplest form and usually should aim to present but one idea, they should be well drawn and the wording should be clear and large enough to be easily read. If charts are going to serve their various suggested ends effectively, attention must be paid to selection and such details of construction.

Graphic advertising is too recent a development to enable one to assemble much information as to the results obtained.   Most of the examples are special cases and cannot be isolated from the general campaign the concern may be conducting.   There are as yet but few instances of a campaign which centered around a series of charts.   There were a number of cases where charts were used repeatedly, however, in various ways as a part of the campaign: as for example, the Thermoid stoppage chart and the Tydol "What explodes in your engine."   The Byers Pipe advertisements are holding to the one idea of 5 per cent. increase in cost and 100 per cent. increase in life and employing various graphic devices to get it over.

The concerns which were using graphics consistently were asked for information as to results secured.   The answers were uniformly favorable though rather general; for example, the A. M. Byers Company report: "The results have been so good that we are continuing the use of this same style of advertising display during the present year." The Thermoid Rubber Company, speaking of the stoppage chart says: "The graphic form of this chart has been found very successful. It is used throughout our entire brake lining campaign."

CHAPTER XXIII

GRAPHIC CHARTS IN THE COLLECTION DEPARTMENT

It is the duty of the collection manager to see that the accounts due his company are paid promptly in accordance with the terms agreed upon.

The collection department should work hand in hand with the credit department. In fact, in many concerns the duties of both are embodied under one head. It is essential that the collection manager

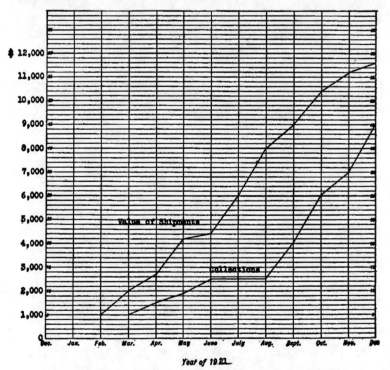

Fig. 1—Graphic Chart Showing Value of Shipments and Collections

know the financial standing of the customer as determined by the credit manager and it is equally important that the credit manager know whether the customer lives up to the terms he has made with the company. See Chapter XXV.

Graphic charts will enable the collection manager to visualize the status of his accounts, both individual and collective, and hence to keep in closer touch with them, better than any other method.

Fig. 1 is a chart showing the Value of Shipments to a customer compared with the Collections received from him. If the curves are parallel, or nearly so, it indicates that the customer is living up to his obligations. If the collections drop off, it is clearly shown on the chart. The average time it takes the customer to pay is indicated by the average "lag" of the Collections curve behind the Value of Shipments. Fig. 1 is a cumulative chart, and it shows that at the outset the customer

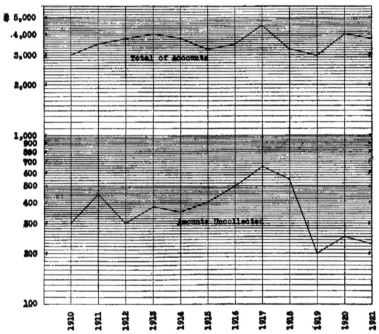

Fig. 2—Graphic Chart Showing Total of Accounts to be Collected and Amounts Uncollected

paid one month after shipment. His payments gradually lagged until in June they were over two months behind shipment. In July and August he paid nothing, but from September to December he paid at about the same rate as he received and there was a lag of about three months.

Fig. 2 shows the efficiency of the collection department over a period of years. The upper curve on the sheet is the total amount of accounts turned over each year to the department for collection, and the lower curve is the amount uncollected. If the amount uncollected increases at a faster rate than the total of accounts increases there is probably something wrong with the collection department. The fault may, however, be due to the credit manager in advancing credit to people who are slow in paying. With the aid of the Component Parts Scale, described in Chapter XI, the percentage of "uncollectables" to total may be read off directly from month to month.

Plain vs. Ratio Ruling. The following is taken from an article which appeared in a recent magazine. The idea suggested is excellent but we believe that the data would show to better advantage on ratio ruling, and have redrawn the chart. "Fig. 3 affords a check on the Collection Department. Each block represents the value of "slow pay" accounts turned over to the department for collection during the year. The solid black portion at the top of each block shows the amount uncollected or declared "uncollectable" at the end of the year. It is reasonable to expect that 1917, with a larger volume than 1916, should show a certain increase in uncollectable accounts. In 1918, the chart shows that a much smaller volume of accounts were turned over for collection, and yet nearly as much remained at the end of 1918 as remained in 1917. As there was no apparent reason for this, something must have been the matter with the department—possibly they were loafing on the job. At

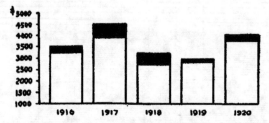

Fig. 3—Chart Showing Efficiency of Collection Department

all events, the general manager, on examining the charts for previous years, saw immediately that something was amiss. He went into the matter carefully and discovered the trouble with the result that in 1919, as the chart shows, the staff made a much better showing, which has evidently continued during 1920."

The chart shown in Fig. 3 is misleading because the scale starts at 1,000 instead of at zero. The black portions represent the amount of the accounts declared uncollectable and are directly compared (by area) with the white portions. As the white portions are cut off because the scale does not go to zero, a wrong idea of the relative value of the uncollected amounts to the total collections is given. We have redrawn the chart, Fig. 4. It will be noted that this chart shows to better advantage the relation of the two items.

The article says, "In 1918 the chart shows that a much smaller volume of accounts were turned over for collection, and yet nearly as much remained uncollected at the end of 1918 as remained in 1917."

This statement is correct, but it should be percentages and not amounts that concern the collection department. Presumably there will always be a certain percentage of the accounts turned over for collection which it is impossible to collect, and this percentage should not vary appreciably from year to year. When drawn on plain ruling it is impossible to obtain a correct picture of this percentage and therefore impossible to tell whether it is increasing or decreasing. Fig. 5 shows the data recharted on ratio ruling. It will be noted that the solid black areas bear quite a different relation to one another on this chart than

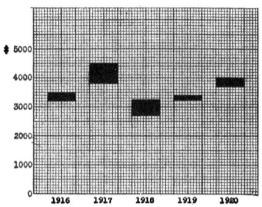

Fig. 4—Fig. 3 Redrawn to Show Zero

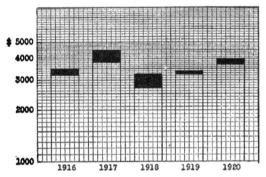

Fig. 5—Fig. 3 Redrawn on Ratio Ruling

on plain ruling, Fig. 3 and 4. This is especially noticeable for 1917.

A true picture is given by the ratio ruling and it is recommended that this ruling be used for studies of this kind.

Keeping Tabs on Collections. The following, from an article in *System, the Magazine of Business,* June, 1918, ''Getting Your Money When It's Due,'' illustrates the value of the chart in this connection. Mr. Horace S. Griswold, who uses the chart, says: (See Fig. 6).

Here are plotted cumulatively, from month to month, sales and accounts received; using the accounts received not of the same date as the sales, but two months later. January sales are compared with March collections. January plus February sales are compared with March plus April collections. The same relation holds throughout. Working on a 60-day basis if accounts are collected up to date, the accounts-collected line should be nearly parallel with the sales line.

Seasonal lags and spurts in collections are thus clearly shown, and the general relations between collections and sales can be properly

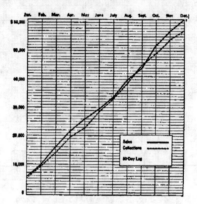

Fig. 6—How Collections and Sales
Agree

maintained. The objection might be offered by some that moneys collected during March might apply on sales in March or any prior month. That is true, but in spite of some hits and some misses the general relation holds nearly enough true to make the chart effective.

I find that the use of the graph for general conditions, and graph cards for individual variations, serves to warn me of dangerous conditions and tends to keep collections where they should be.

CHAPTER XXIV

Graphic Charts for Showing Costs

There is nothing equal to the graphic chart for showing costs, whether it be for the purpose of studying the fluctuations in the cost of operating a machine, a department, or an entire business, or whether it be to compare actual cost of performance with some predetermined standard or "bogey."

Charts for studying fluctuations in operating costs may be of any of the forms previously described—bar charts, line charts, circular percentage charts, etc. Their object is to reduce to graphic form records of costs extending over any desired period. The charts show up the variations as nothing else can and furnish a basis for analysis and study which it would be difficult, if not impossible, to obtain in any other way.

There is almost no limit to the use of charts in this connection and it is safe to say that they are used more extensively for the study and analysis of costs today than for any other purpose. It is suggested that every business man who reads this book make up a chart showing the costs in some department of his business over a period of six months or a year. It is almost certain that he will discover therefrom the existence of certain conditions which he has not previously suspected.

The use of charts to compare results with set standards was described by W. N. Conner in *System*, May, 1920. A portion of his article follows:

Recently we have extended the use of our graphics beyond the executive activities. And there too—particularly where it is possible to inculcate rivalry among different workmen and different foremen—graphs and charts have helped us materially to speed up production or to lower costs, or both.

We have charts plotted week by week for the information of the foremen and their men on the construction jobs. These charts show the unit costs of each class of work—making up forms, for instance, erecting forms, erecting steel, placing reinforcement, pouring concrete and the like.

Incentives are lacking to lower costs, we have found, when there isn't a mark to shoot at. So, near the beginning of each job, we set a so-called "bogey." This bogey is the figure which past experience shows us we may reasonably expect as an average unit labor-cost for the job.

The cost curve as plotted week by week may go above or below the bogey, according to the progress made in the work and the way the costs are handled.

No matter how complete may be the information tabulated, charted and placed before the executives of a company, as long as this information goes no further down the line it furnishes neither the required incentive nor the definite information to the men on the job, as do our simple graphs. And it must be remembered that it is the men on the job who are actually making the costs by their daily work.

Only by relating the costs on a new job to those on previous jobs on which the same men have worked, or to those of other parts of the same

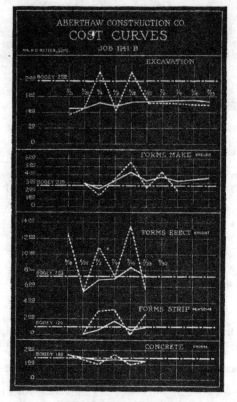

Fig. 1—Cost Curves

job where rivalry can be set up, can the highest interest of the men themselves—in our experience—be aroused. Consequently, we have devoted much effort to teaching the foremen, carpenters, masons and other artisans what these cost curves mean. We have succeeded in instilling in them a desire to talk the costs over among themselves, and keep them as low as is consistent with first-class work. This we have been achieving to a very considerable degree for some little time.

The cost man—we keep such a man in the construction office on each job—works out the figures every week. He posts the costs in a weather-proof case on the outside wall of the field office. And at the same time, he posts a chart on which he has traced the results of the past week's work. In many instances the interest which the charts and figures have aroused among the men has been so great that almost every worker has followed them carefully from week to week with unusual closeness.

The chart, Fig. 1, shows typical cost curves; it is the chart, with its 5 series of graphs, showing the costs—and posted in the construction office—of a large job near Utica. The chart shows the bogey as a heavy dot-and-dash line running across the diagram on each of the separate parts of the job. The weekly figures for each operation are shown by a heavy line of short dashes, fluctuating considerably week by week on most of the jobs.

The total unit cost of the operation from the beginning of work and carried right up to the end of each week is shown in the heavy

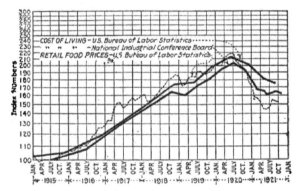

Fig. 2—The Trend of the Cost of Living

solid line, which fluctuates less rapidly than the weekly costs. Since it is the average of the weekly costs to date, it of course tends to rise when the weekly figure is above and fall when the weekly figure falls below. As the weeks stretch out into months, it is evident that this over-all figure will become more and more uniform, and ultimately will show very little fluctuation, except when there happens to be excessive variations in the weekly figures.

Graphic Chart of the Cost of Living. The following is taken from the December, 1921, issue of Management Engineering.

The cost of living in the United States during October was practically the same as in September, according to the figures gathered by the National Industrial Conference Board. Fig. 2 shows the changes which have taken place in the cost of living since January, 1915. Living costs it will be noted are still approximately 64 per cent. above pre-war levels.

CHAPTER XXV

GRAPHIC CHARTS FOR THE CREDIT MANAGER

The credit manager will find graphic charts a valuable asset in properly managing his department. As he is responsible for the extension of credit to the customers of the company he must be in close touch with their financial condition. Where large amounts of credit are extended the credit manager should have copies of the balance sheets of the customer. These will furnish the data from which he may make a number of graphic charts to assist him in his work.

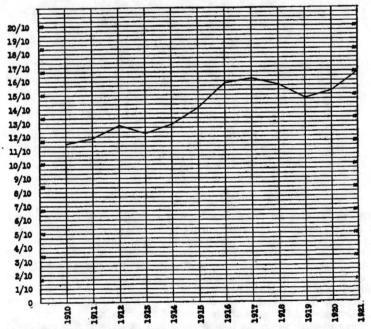

Fig. 1—Graphic Chart Showing Relation of Liquid Assets to Current Liabilities

The ratio of liquid assets to current liabilities is one of the main factors in deciding whether or not to extend credit, and how much to extend. If the percentage is high, the prospective customer may be considered in a favorable light. The graphic chart, Fig. 1, shows this ratio over a period of years. The increasing ratio shown by the upward trend of the curve indicates that the customer is a sound risk and

134

that the amount of credit extended to him may be safely increased, with the trend of the curve as a guide. If the trend of the curve shows a downward tendency, it is apparent that the condition of the customer's business bears close watching.

The credit manager is, of course, interested in knowing whether the customer is making money, for it is to the advantage of his company to do as large a business with the customer as is safely possible. Fig. 2 shows a chart on which the net worth of the customer, as given by

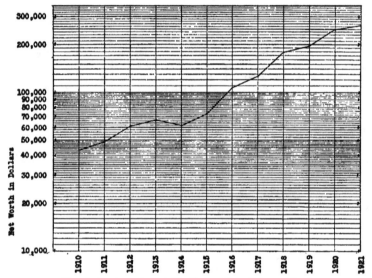

Fig. 2—Customer's Net Worth from his Balance Sheets

his balance sheets, has been plotted over a period of years. If the customer is making money, the trend of the curve will be upward; if he is losing money, it will be downward. Ratio ruling should be used for a chart of this type. If plotted on plain ruling, the chart is very liable to be misleading and may cause, or influence, the credit manager to make an unwise decision.

The cumulative graphic chart, Fig. 1, Chapter XXIII, compares the value of shipments to, and collections from, the customer. A satisfactory condition is shown if the curves are parallel, or nearly so, and if the lag is not too great and does not tend to get greater. If the value of the shipments shows an increase, and the collections drop off, it will be clearly shown on the chart, and under these conditions no time should be lost in revising the terms of credit.

CHAPTER XXVI

Graphic Charts to Show Financial Facts

In his article, "Statistics in Business," published in *Administration*, May, 1921, Walter B. Cokell says: The field of financial statistics is perhaps the broadest of all fields, and data of this character are used daily by most executives. Mercantile agencies issue weekly statements of market conditions which are supported by statistics of the volume of trade, prices of commodities, business failures, pig-iron production, building permits, and the like. Wall Street operators and financial publishers issue daily or weekly summaries of stock-market conditions. In these are found statistics of the sales and prices of stocks and bonds, money rates, bank clearings, railroad earnings, foreign trade, and all figures of business activity which are of value to an executive. Many banking and brokerage institutions also issue economic analyses of trade conditions in general, with more or less detailed studies of activities in various industries. The backbone of some, if not most, of these analyses is statistics, of which the main part of the reading matter is the interpretation thereof.

Some of the larger corporations compile monthly reports of business conditions for distribution within their own organizations. The idea usually is to summarize business conditions in general, compare these with the results of operations of the corporation itself, and then interpret the correlation of the two.

In the financing of a business most executives find a summary sheet of revenues, expenses, and net revenue, such as illustrated in Fig. 1, very useful in obtaining a bird's-eye view of the situation. From the study of this chart it is evident that the business is greatly affected by seasonal fluctuations. Revenues drop off in the summer much faster than expenses can be reduced, consequently the net falls even more. In August expenses increase before revenues grow, probably because of large advertising expenditures, overhauling of plant, etc., in preparation for the fall trade.

An interesting use of charts in interpreting the business situation is given in an article by F. Leslie Hayford in the April, 1920, issue of the Du Pont Magazine. While the following material is based on the situation in the first part of 1920, the general method of approaching the subject may be used at any time. The following notes and Fig. 2 are from Mr. Hayford's article.

Six statistical series or groups of series have been selected for consideration in this article. They are: (1) wholesale price index numbers, which show changes in the general price level; (2) pig iron production for the entire country and unfilled orders of the U. S. Steel Corporation, which indicate the condition of the iron and steel industry; (3) bank clearings, which are the best indices of the volume of business transacted, particularly clearings of banks outside of New York City: (4) exports and imports, which show the extent and general nature of our trade with other countries; (5) the number of business failures which indicates whether business is prosperous or unremunerative; (6) the

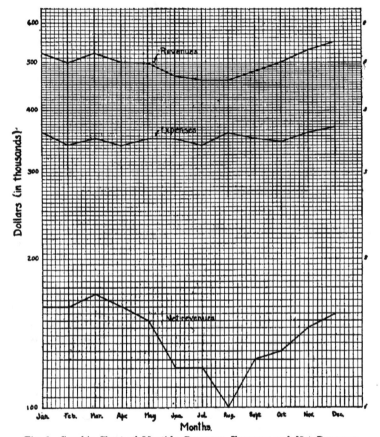

Fig. 1—Graphic Chart of Monthly Revenues, Expenses and Net Revenues

value of building permits for twenty leading cities, this being the best index of building activity. These series are shown in graphic form in Fig. 2.

It is not assumed that these statistical series, as they are here presented, are exact barometers of business conditions, for some of them are subject to fluctuations which tend to obscure their significance. For example, as the population of the country grows, business also tends to increase in volume; thus a long-time tendency (or growth)

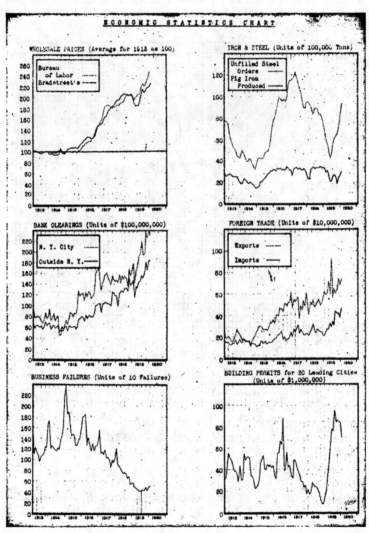

Fig. 2.

is apparent in some of these series. Seasonal variations are also apparent, particularly in building permits, which usually are largest in the spring. Scientific methods of measuring these long-time tendencies and seasonal variations have been worked out by the Harvard University Committee on Economic Research. It is thought, however, that the graphic presentation of the actual figures will prove of interest to the readers of the Du Pont Magazine. But it must be recognized that they are only rough indices of business conditions and not exact means of measurement.

Wholesale Prices. Two wholesale price indices are shown. In each case the average of monthly prices for 1913 is taken as the base, or 100, and the data for subsequent months are shown as percentages of this figure. During January prices continued to advance, reaching the highest point ever attained. The Bureau of Labor Statistics index number (based on 328 quotations) for that month was 248, indicating an increase of 148 per cent. in general wholesale prices since 1913. Bradstreet's index number (based on 96 quotations) for the first of February was 227, showing an increase of 127 per cent. since 1913. This disparity between the two index numbers is doubtless due to the difference in the number of commodities upon which they are based and also perhaps to the differences in methods of computation and dates for which price quotations are taken.

Iron and Steel. War conditions apparently altered the relation of the iron industry to general business, and the statistics of pig iron production and unfilled steel orders may not yet have regained the significance which they formerly had. The unfilled orders of the United States Steel Corporation have shown a steady increase since last May. Pig iron production was apparently much hampered by the strike last fall, but since October it has been increasing. January figures of unfilled steel orders were 9,285,441 tons, a larger unfilled tonnage than at any time prior to the war, although less than during the height of the war boom. Pig iron produced during January amounted to 3,915,181 tons.

Bank Clearings. New York City clearings are largely affected by stock market operations and, therefore, should be studied in connection with stock market movements. Clearings for the United States outside of New York are probably the best single index of business activity available. Bank clearings should be interpreted in connection with the wholesale price indices, as they reflect changes in the price level as well as changes in volume of business. In this connection it is interesting to note that clearings outside of New York amounted to 14½ billion dollars in January, 1919, and 18 billion dollars in January, 1920, an increase of approximately 24 per cent. During the same period, however, general wholesale prices increased approximately 22 per cent., according to the price index of the Bureau of Labor Statistics.

Foreign Trade. In December both exports and imports fell off slightly, but both showed improvement in January. In the case of imports the January figures established a new high record, amounting to $477,000,000 an increase of 96 million dollars (or 25 per cent.) over the preceding month. Exports during January totaled $731,000,000. In interpreting statistics of exports and imports, it must be borne in mind that they are in terms of dollars and that their significance is modified by changes in the price level.

Business Failures. The graph shows the number of failures each

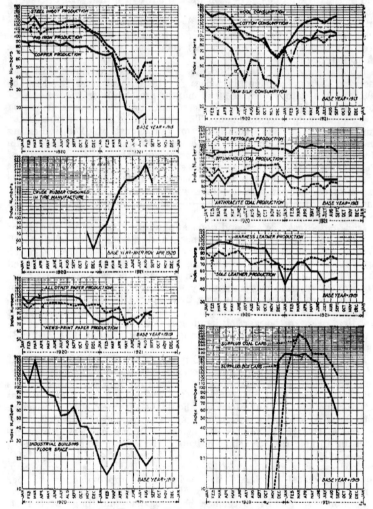

Fig. 3—Trends in Industry Since January, 1920

month in the United States as reported by *Bradstreet's*. During the past year the number of failures has been unusually small. It is perhaps a fair interpretation to say that this fact indicates that high prices and extravagant spending have made it possible for many inefficiently managed businesses to get along, and that any pronounced lessening of business activity might easily produce a much-increased crop of failures. In January the number of failures was 511, compared with 488 the preceding month and 573 the year before.

Building Permits. The combined valuation of building permits

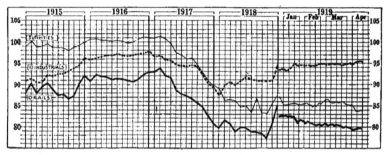

Fig. 4—Price Trend of New York Bond Market

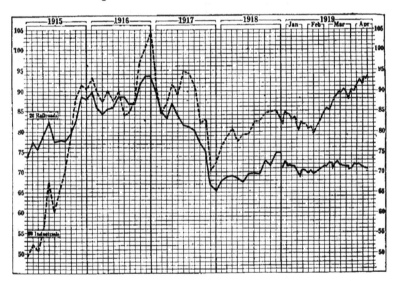

Fig. 5—Price Trend of New York Stock Market

issued each month in twenty of the leading cities of the United States is shown in the graph. The estimated value of buildings to be erected is represented and not the amount of contracts actually let. The effect of war-time restrictions is apparent in the fall of the valuation of permits to 8½ million dollars in December, 1918, the lowest point reached during the seventeen years for which comparable figures are available. With the removal of such restrictions, however, the great need for new buildings began to make itself felt, with the result that the monthly value of building permits for these twenty cities rose to 95 million dollars in August, 1919, and exceeded 80 million during each of the four months following. In January building permits amounted to 70 million dollars.

Summary. Prices have reached the highest level ever attained; the iron and steel industry is active; bank clearings indicate only a fair volume of business instead of a great increase over a year ago; exports still hold up in spite of low exchange rates, while imports are on the increase; business failures are very few in number; building permits indicate great building activity during the next few months.

Analysis of other economic data shows that tightness exists in the money market, while the demand for capital is great. Some decline in prices is indicated, as well as a lessening in business activity. But inasmuch as no real surplus of commodities seems to exist, a long-continued depression is not anticipated and business should make a rapid recovery.

Industrial Index Numbers. The United States Department of Commerce publishes a *Monthly Survey of Current Business*, the present subscription price of which is $1.00 per year. This survey contains tables and charts showing business trends, in most cases from 1913 to date. It covers the movements in industry, agriculture, foreign trade, banking, wages, etc., and the data are collected from federal and state government reports, trade associations, private organizations and reports from technical periodicals. It thus gathers under one cover much material which heretofore has been obtainable only from a variety of sources.

Many of these data are invaluable for the executive and may be presented on graphic charts in the most concise and quickly available form. Most of them should be charted on ratio ruling to be of maximum value.

Fig. 3 is reproduced from the December, 1921, issue of *Management Engineering*. The magazine states that the charts were drawn from the data of the *Monthly Survey of Current Business*. In most instances the year 1913 is taken as 100 and the relative variations computed upon that basis.

Trend of Bond and Stock Markets. Figs. 4 and 5 are examples of charts published from time to time by the financial department of the New York *Tribune* to show the course of the bond and stock markets in New York over a considerable period. The use of ratio ruling would give a much better idea of the relative percentage of increase and decrease.

CHAPTER XXVII

GRAPHIC CHARTS FOR INVENTORIES

Graphic charts furnish an excellent means for keeping a perpetual inventory. They will show at a glance the amount (or value) of material on hand, how fast and when the material leaves stock, when the "danger point" is reached and it is time to order new material when the new material is received and how much is received, etc.

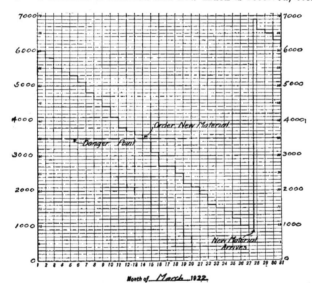

Fig. 1—Perpetual Inventory Kept by Graphic Chart

Fig. 1 is a simple illustration. On March 1 there were 6000 units of material on hand. From March 1 to 14 the material went out at the rate of about 200 units per day. On March 14 the "danger point" was crossed and an order put in for new material. This new material was added to stock on the 28th. There were 6000 new units and the stock was raised from 900 to 6900. 300 units went out on the 28th and a certain number each day thereafter as shown.

Of course any period of time—week, month or year—may be covered and by whatever intervals—days, weeks or months—are desired. Any other data may be added to the sheet, such as the order or job numbers, destination of material, unit cost of material, etc.

Some of the advantages of this method are: (1) Ease of keeping the record. (2) The great amount of information given. (3) Guards against delays due to waiting for material, as the approach to the "danger point" is instantly perceived. (4) By checking up with the actual count of the material at intervals, any serious differences may be investigated to the end that the stock room be better protected. (5) The value of the stock on hand may be quickly ascertained at any time.

CHAPTER XXVIII

THE ORGANIZATION CHART

Organization is the mechanical framework or substructure through which the functions of management operate.

There are various types of organizations in use but, regardless of the form, there should be no doubt in the minds of those connected with an organization as to how it operates.

It is to give a clear understanding of the workings of the organization

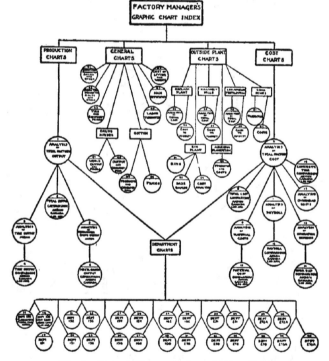

Fig. 1—Factory Manager's Organization Chart
Goodyear Tire & Rubber Co.

—to define the authority of the various heads and sub-heads, and to illustrate just exactly what the functions and responsibilities of each member of the organization are, that the organization chart is useful.

145

It gives a picture of the layout of the organization as a whole and shows the relation to one another of the various units that go to make it up. Fig. I illustrates this type of chart. It shows at the same time what an important part charts play in the organization of the Goodyear Tire & Rubber Co.

L. V. Estes, in *Industrial Management*, January, 1920, gives the following on the part the graphic chart plays in organization.

One of the problems before any president or general manager who is anxious to have control of the business in his charge is to find means for binding his organization together into a whole that will function without friction, and be amenable to his leadership. The first step toward getting control of a business and its production is to see that it is properly organized and that the persons who fill the various posi-

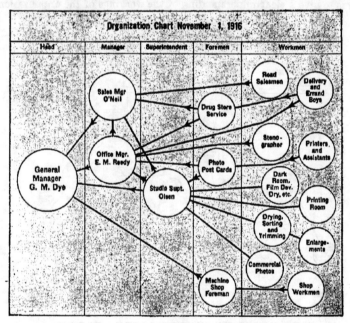

Fig. 2

tions in the organization understand what their relationship is to each and every position of the organization. This latter object may most easily be obtained by charting the organization. The basis of control is adequate organization thoroughly understood by its members.

An Application of the Organization Chart. The following is from an article by G. M. Dye in *System*, December, 1919.

Our business deals with photography. We develop and print films, variously called amateur finishing, or photo-finishing.

Our customers comprise some 600 drug stores in the vicinity of

Minneapolis, and mail-order clients extending as far as Montana. And that's just one part of the business.

The other part is the manufacture and marketing of improved picture-finishing machinery for many other finishing concerns like ourselves.

It requires 50 machinists to take care of this division and 80 more employees for the photo-finishing department.

At first glance it may seem a bit awkward to handle a picture-finishing department in connection with a machine shop. In reality they are interdependent. You see we cannot afford to do our finishing except by most improved machinery. So we manufacture accord-

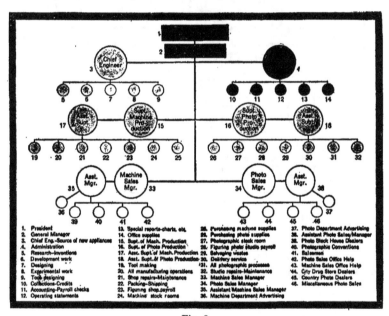

Fig. 3

ing to our needs and give each new machine a thorough tryout before placing it upon the market.

In a business as complicated as ours, charts have enabled us to detect the weak spots in the organization, and strengthen them in a way utterly impossible without such visualization.

Suggestions gleaned from reading have shown me the possibilities in the use of charts in building up a personnel. Three years ago I drew our first organization chart, reproduced in Fig. 2. Since then, with us a revised chart has become an annual affair. The important point is that each revision has been one step nearer the diagram best suited to our ultimate needs. Each shift in our personnel is a move towards fitting the men to the diagram rather than being obliged, as at first, to fit the chart to the men.

Our ultimate organization chart, as we now see our future, is shown in Fig. 3.

Fig. 1 is made up of a combination of rectangular "boxes" and circles. More frequently only one kind or the other is used. Where

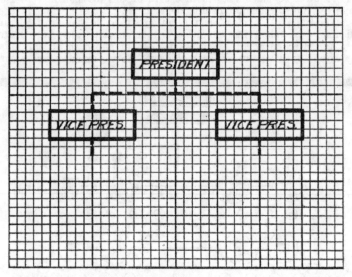

Fig. 4—Cross Section Paper, 10 Divisions per Inch, Handy for Drawing
Organization Charts

the "boxes" are used, cross-section paper like that illustrated in Fig. 4 is very handy as the lines furnish guides both for drawing the boxes and for lettering in the desired names, etc.

CHAPTER XXIX

Graphic Charts for Payment Plans

There are in use at present a great many different systems of compensation for workmen, varying from the oldest and probably the most extensively used form—daywork, to the various forms of bonus and efficiency systems. A description of these will be found in any good book on management engineering and it is the purpose of this chapter merely to state that graphic charts may be used to excellent advantage in connection with the planning and operation of these various systems and to show, by a few examples, how they have been employed in various ways.

Bonus and Piecework Rates. Charts may be used to great advantage in laying out bonus and piece-rate curves, not only for figuring what amounts shall be paid but also as an aid in determining a uniform and consistent rate of payment.

Fig. 1 is from an article by C. E. Knoeppel, "The Bonus Plan of Wage Payments," published in *Engineering Magazine*, November, 1914. Mr. Knoeppel says: Fig. 1 will show the bonus curve used and advocated by me. The heavy curved line is the bonus line starting at 67 per cent. efficiency, which means the workman is expected to attain two-thirds of a fair standard before he begins to earn anything additional in bonus. In other words, the man can take 50 per cent. more time than that called for by the standard, for which he receives day wages only. Any reduction in time under this 50 per cent. margin would be accomplished by a proportionate amount of bonus.

The bonus line is divided into three sections, A, B and C. Men of low efficiency do not become men of high efficiency over-night. They sometimes feel that they cannot attain the standard determined upon. The aim is therefore to induce men to work up the "A" incline from 67 per cent. to 85 per cent. efficiency. They then have the incline "B" ahead of them, with promise of additional earnings if they get into this class. Men are not satisfied with standing still, nor do they want to be considered as low-efficiency men. Finally when men are well along towards the 100 per cent. mark they are attracted by the additional 5 per cent. premium for qualifying as 100 per cent. men. A bonus of 20 per cent. plus 5 per cent. premium seems worth the additional effort to the man who is within 3 per cent. or 4 per cent. of the goal.

For comparative purposes the Emerson curve has been shown by a dotted line where it varies from the curve recommended. The extra amount indicated by the shaded zone 1 is to warrant men, who might otherwise become discouraged in making the effort necessary, in attempting the attainment of efficiencies greater than 67 per cent., and the amount measured by the shaded zone 2 is a premium for those who average 100 per cent. of the standards or better.

The chart in question shows a third line (dot and dash) which may be interesting to the student of bonus plans. The claim has been made that because the Emerson and Knoeppel bonus lines mean slightly decreasing costs per piece, they are unfair to the workmen; that the rate per piece should remain constant as in the straight piece work plan. Bonus paid on the basis of the "X" line does this, and its comparison with the other two lines will be found interesting.

Mr. Knoeppel, at some length, interestingly and thoroughly develops his various principles for bonus payment and says:

The matter of an intelligent and comprehensive control of the entire work is most important. To take care of this feature properly a number of charts can be used to decided advantage.

Fig. 2 is a record of the bonus earned per man per period. I was once bitterly accused of being too anxious for the men to earn bonus. *I am.* When men earn bonus it means that efficiency, and therefore

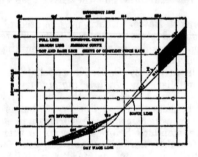

Fig. 1—Bonus Curves

production, is higher than if they were earning no bonus. It is a good plan to know what the standard earnings should be, which on the chart are shown by a dotted line. The actual bonus earnings come within 20 cts. of the standard period ending 6/28. The chart shows a healthy condition in that bonus earnings per man show a steady increase.

In order to keep in close touch with the progress of bonus men, the chart shown in Fig. 3 is suggested. Two things are essential—

(1) There should be a steady increase in the number of men put on bonus.

(2) Those on bonus should have as much of their time covered by schedules as is possible.

On the chart the heavy line shows the ratio of the time of bonus men on schedules to the total time they work, while the dotted line shows the ratio of time of bonus men on schedules to the time of all men in the department or plant. Take the period ending 5/24 for example; more men were put on bonus, but the time on schedules was less than in the previous period. The dotted line for period 5/31 shows a falling off in the number of men on bonus, although those who were on worked on them 65 per cent. of their time. The heavy line for period 4/26 and 5/3 shows such decided drops as to warrant rigid investigation. Both lines, however, show an upward tendency, which is, of course, encouraging.

The "inefficiency chart," Fig. 4, is decidedly necessary. *My claim*

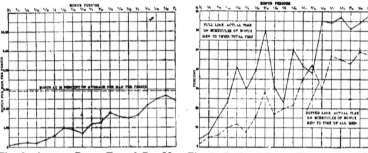

Fig. 2—Average Bonus Earned Per Man   Fig. 3—Ratio of Bonus Time to Possible
Per Period   Time

is and has been that inefficiency is the element to analyze, for we increase efficiency only through eliminating inefficiency. Further, the inefficiency of management should be shown as distinct from that of the men. If this is not done there can be no true conception of what is at fault and who to blame. This is accomplished by adding the allowances to the actual hours, after the man-efficiency for a department has been determined, and dividing the same figure for standard hours

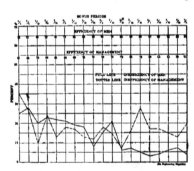 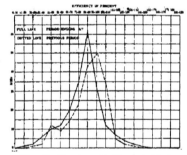

Fig. 4—Inefficiency Chart   Fig. 5—Number of Men at Various
Efficiencies

that was used in figuring the man efficiency, by the increased divisor in the form of actual hours. The full line, or man inefficiency, shows a constant decrease although in period 5/10 it increased noticeably. This increase might be due to putting new men on bonus, or old bonus men on new work, or other causes, like cutting a rate or arousing the opposition of the men. From period 4/26 both inefficiency of management and men increased after several periods of decreases. Then comes the sudden drop in both for period 5/17. The significant fact in this connection is that following the period 5/17 the man inefficiency decreased while *the inefficiency of the management took two upward spurts.* Further the lines show that the men are making faster progress in eliminating their inefficiency than the management, the moral of which is—*"get after the management."*

Fig. 5 is important in showing the number of men at classified efficiencies. For the period in question the chart shows that 127 men attained efficiencies varying from 71 per cent. to 100 per cent., while in the previous period only 116 men attained these efficiencies. At the same time the general showing for the previous period is better than for this period, in that there were less men showing efficiencies from 51 per cent. to 90 per cent. and more men from 91 per cent. to 110 per cent.

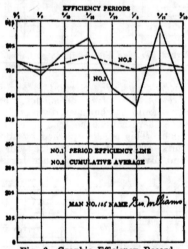

Fig. 6—Graphic Efficiency Record

The value of the chart lies in the ability to concentrate attention on the men showing efficiencies of 80 per cent. and under, and ascertaining what interferes with their attaining greater efficiencies. Further, this chart is especially valuable in connection with Fig. 3 showing relative times. *The work of getting more men on schedules, keeping those who are thus employed on schedules for the greatest part of their time, and getting the men showing one class of efficiency into the next higher, can be planned from these two charts.*

To assist further in the work of eliminating inefficiency a sheet should be prepared covering the efficiency of the workers for a period and posted in a place where it can easily be seen, Fig. 6.

As Soon as a Payment Plan has been Installed the management wants to know the result, just how much wages and production have increased and how much the unit cost of production has decreased. The best method of presenting these results is by a graphic chart. Fig. 7 shows what followed the installation of a bonus plan in a large creamery. The time studies, determining of standard times, devising the bonus plan, etc., were done in two months and resulted in increased production, in stabilizing the hours of working and in a decided increase in wages. The bonus plan is described in detail by E. N. Hay in November, 1917, *Industrial Management*. Mr. Hay says, in part:

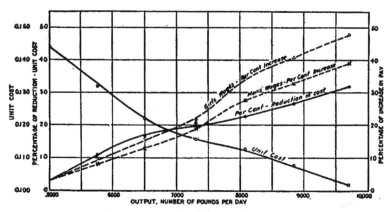

Fig. 7—Results of Bonus Plan Installed in Creamery

At the beginning of the week a chart sheet, Fig. 8, was posted in each department to show graphically the performance of each gang in that department. This performance or efficiency was expressed by the ratio between two lines, one a vertical line showing the work that should be done in the number of hours worked by that gang and beside it a line representing the work actually done. The ratio of the heights of these two lines gave the percentage of efficiency of that department for the previous day and this efficiency was written down in per cent. at the foot of these lines. The bonus earned for that day was also set down in dollars and cents so that each gang could see just how much bonus it had earned on any given day of the week. At the end of the week the total bonus was given in cash in an extra envelope.

Another Very Valuable Use of graphic charts in connection with wage payments is for the calculation of the payroll, of the percentage of actual output to standard output, etc. Such a chart is described in the August 15, 1920, issue of *Factory* under the title, "Letting a Chart Do the Figuring." It was used at the plant of the H. Mueller Mfg. Co. The article says, in part:

This particular chart, Fig. 9, was designed to help determine standard hours credit, and we use it in connection with production control boards and bonus calculations. It may be used also for determining standard quality or actual quantity per hour or per unit of time. Nearly every executive will see other uses for such a chart where three quantities depend upon one another and which can be adapted and slightly changed so it will fit his particular needs.

The three quantities represented on this chart are: actual quantity produced, shown on the horizontal scale; standard hours of credit, indicated by the vertical scale; and standard quantity, which is represented by the slanting lines.

The marking of values on the scales is indicated on the chart. The chart which we use is several times the size of the one illustrated, and has vertical subdivisions one-fifth the size of the unit chosen.

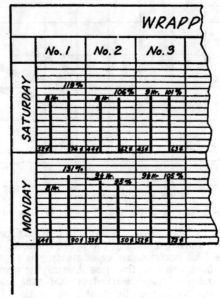

Girls' pay given in cents at left side and men's at
right side for each gang

Fig. 8—Section of the Daily Efficiency Chart

Of course, the figures used on the chart may represent the same quantities as shown, or those quantities multiplied by 10 or any power of 10. For example, line 4 may represent .04 or 40,000, or any other product of 4 and a power of 10. It is necessary only to determine mentally the decimal point to make these scales represent quantities of any magnitude. The use of the charts can best be represented by some actual problems. Suppose the standard quantity is 1,000 per hour and the actual quantity produced is 600 per hour. If we want to find the standard hours credit we will follow the vertical line 6, representing actual quantity 600, to the intersection with slanting line 10, representing standard quantity of 1,000. The horizontal line which passes through this point is 6, but by inspection it is seen that the result should have a decimal point in front of it, so the result is taken as .6 in standard hours credit.

Such a simple problem, of course, could be more easily solved mentally, but when the figures are more complicated the chart gives the most rapid solutions.

Suppose the standard quantity were 2, the actual quantity 5. The vertical line 5 does not cross the slanting line 2, so it will be necessary to follow the vertical line .5 until it intersects 2. The result read on the horizontal line is 2.5 standard hours credit.

If the standard quantity per hour is wanted when the standard hours credit and actual quantity per hour are known the method is practically a reversal of the one given for the other problem. Suppose, for ex-

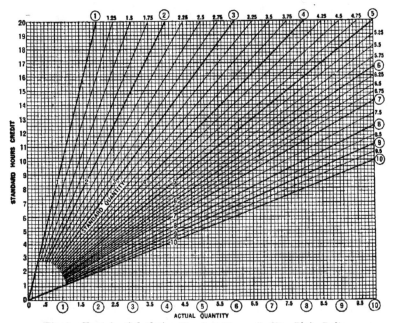

Fig. 9—Chart for Calculating Standard Hours Credit. Plain Ruling

ample, that the standard hours credit is 5 and the actual quantity is 75 per hour.

The horizontal line 5 and the vertical line 7.5 do not intersect within the margin of the slanting lines. But the 5 horizontal line does intersect both 5 and 1.0 within the range desired, so a line midway between these can be assumed and the slanting line passing through the intersection of 5 and .75 will indicate 15 as the standard quantity per hour. In all of these cases, of course, the decimal point is placed by inspection.

If this chart were drawn on logarithmic paper the lines would be parallel instead of converging, but the chart would operate in a similar way. In Fig. 10 the main lines only, 1, 2, 3, 4, etc., have been drawn, in order to show how the chart looks and operates on logarithmic paper, but intermediate lines, 1.25, 1.50, 1.75, etc., may be added as in Fig. 9.

Chart An Aid in Determining Piece Rates. An Experienced Time-Study Man says in *Factory*, June 1, 1920, that when twenty minutes of an operator's time are plotted on a chart showing how that time was spent, it takes only a glance to see how much time is profitably expended and how much is uselessly wasted. Referring to women workers sewing salvage seams on machines he says: The greatest problem confronting the observer is a distinction between productive or necessary

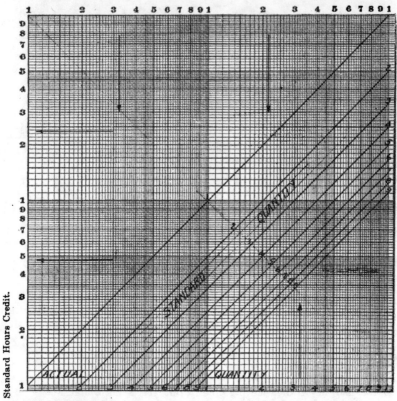

Actual Quantity.

Fig. 10—Chart Similar to Fig. 9 Drawn on Logarithmic Paper

time, and unproductive or wasted time. Time may be wasted in various ways: unnecessary motions, talking, looking around at other operators, and repairing of spoiled work. In order to place this before the mind of the observer and management, a graph similar to Fig. 11 should be made up covering a given interval of time, and showing the time consumed by the various movements of the operator. After the graph is made, the operator can be shown wherein she can eliminate waste motions, and standardize her output.

In Fig. 11 the total productive time is the total of the periods of production and the time spent in adjusting the machine. At first glance it might be thought that the operator should be given credit for the time consumed in ripping mistakes, but it was evident that the work which had to be ripped was preceded by an interval of talking, which, without doubt, was the cause of the mistake.

During the period of 20 minutes the operator sewed 5 seams. As the actual productive time was 71 per cent. of the total, her standard for the period of 20 minutes should be computed on a 100 per cent. basis, or 7.04 seams. By multiplying by 3 we get the number of seams sewed per hour, then by 8 to get the number per day, which, in this case, equals 169 seams per day. Now, having the standard, it is easy to get the piece price. The worker heretofore has been on a day rate of

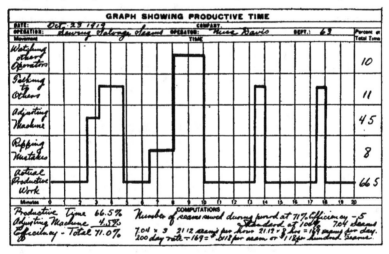

Fig. 11—Chart Showing Results of Time Study

$2. Therefore, we have for the piece-work price $2 divided by 169, giving $.0118 per seam. It is understood that the same workers, now on piecework, will increase their earnings with this incentive, even though the unit price has been figured on 100 per cent. productivity, with no fatigue allowance.

The study of an operation of 20 minutes may not seem a fair average, but as all unproductive time not necessary to the operation is eliminated, it is of no material consequence whether a study of 20 minutes or several hours is taken; the result is the same.

I know of a factory where the entire operations were timed in this manner and a graph drawn up for each operation. By using the same system on each operation there were no abnormal piece-work prices and all operators were satisfied with the standards set. Scientific time study may mean an outlay of money and a breaking up of old routine, but in most cases it more than pays for itself in increased output.

CHAPTER XXX

GRAPHIC CHARTS IN THE PERSONNEL DEPARTMENT

The business and industrial concerns of the country are coming to realize more and more the necessity of keeping records of their employees. The extent to which they go depends upon their size and upon their appreciation of the value of such records. These may vary from a very simple record of the name, address, date of hiring and date of discharging or quitting, which involves part of the time of a single clerk, to complex records of individual characteristics and accomplishments, which require the full time of an entire department.

These individual records are made the basis for the selection, training and promotion of employees and they may also furnish the data for statistics on the number and cause of accidents, the cause and duration of sickness, labor turnover, and many other things. For presenting the results of such studies there is no method as concise, comprehensive and satisfactory as graphic charts.

Graphic Charts as a Means to Visualize Employment Records are described by W. S. Wells in *Industrial Management*, July, 1920. A portion of this article follows: To obtain a correct understanding of working conditions in any plant, full and complete records must first be obtained and made up in sufficient detail to permit careful analysis. For instance, very few executives know what their labor turnover is, or have any idea of the real reasons for it or its effect, nor do they realize the cost in dollars and cents in decreased production and for training labor to replace old experienced employees. Those who have, are giving it careful consideration and study, and are making an effort to reduce it by installing training departments for their new employees in the semi-skilled trades, training their old employees for higher positions, and their foremen to a realization of their proper relation to employer and employee, by means of foremanship courses.

Personnel tells us that there were 41 plants in Detroit whose average turnover for the month of November, 1919, was 19.8 per cent., a yearly average of 237.6 per cent. Only seven of these plants have a yearly turnover of less than 160 per cent. on this basis. Assuming that this average will hold good for the year, every position was filled $2\frac{1}{3}$ times.

Assuming that the cost of labor turnover was $50 per man (which would be a low figure if based upon the investigations made by Mr. Magnus Alexander some years ago), it is costing these 41 firms $1,333,-000 a year for labor turnover, to say nothing of the loss in production and the profits on this lost production.

These charts have been presented with the idea of assisting those who have entered the new field of employment management during the past year or two and have found it difficult to "put across" their policies to the management.

The first step to be taken is to provide a record of the daily activities of the employment department and, for this purpose a labor report is used.

The report is divided into two main divisions, Departmental Changes and Attendance. The first division, "Departmental Changes," is divided into four subdivisions showing the number of men hired, transferred, terminated, and the total number of employees upon the payroll.

For the purpose of comparison and reference with accounting and production reports each of these subdivisions is still further divided under three headings: Administrative, Supervisory and Productive, to show the main divisions of labor following the usual accounting practice.

The totals for departments under the heading "Hired" and "Terminated" are balanced with the number of employees shown under the heading "Payroll," so that this column shows the actual number of employees upon the payroll every day.

The second main division, "Attendance," is divided into non-productive and productive labor, to show whether a proper balance is maintained between these two at all times, as the non-productive labor may easily become excessive if the productive attendance is very low.

A comparison of the total attendance to the payroll column shows the relative working efficiency of each individual department, and the totals the force as a whole.

From these reports the graphs which follow are made up.

The stability chart shown in Fig. 1, shows the fluctuations of the labor force from day to day through the employment department, the effect of these fluctuations upon the total working force, attendance and the amount of absenteeism, and the effect of extreme weather conditions for a calendar month.

The first line at the top of the sheet shows the number of employees upon the payroll, the second line the number of men working each day, and the difference between these two lines indicates the amount of absenteeism.

The next group of lines shows the number of applications, placements, transfers and terminations, and following these a record of the temperature and weather conditions. The greater portion of the men employed in shipbuilding must by the nature of the work be exposed to the weather, therefore extreme weather conditions have considerable effect upon the working force, and it is for this reason that it is important they be shown.

The average number of employees on the payroll for each week is shown by the figures above the line which shows the total number of employees on the payroll. The percentage of absenteeism for the week is shown immediately below this line, while still lower down is given the weekly average working force and the percentage of turnover. The average of these four items for the month follows the last weekly average at the extreme right of the graph.

These charts are summarized for the year on the stability chart shown in Fig. 2, the figures for each month being shown in a table which appears just below the chart.

The changes from day to day are covered by the first chart and are transferred at the end of each month to the second, so that reference

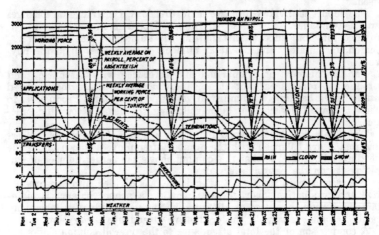

Fig. 1—Stability Chart for the Month of December, 1919

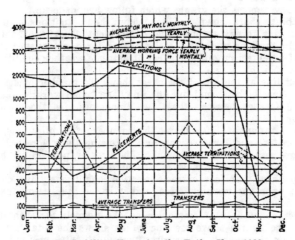

Fig. 2—Stability Chart for the Entire Year 1919

to the two permits the labor conditions being easily followed at all times.

When it is possible to determine in advance the amount of production per year, as is very often the case in manufacturing plants, the estimated or desired amount of production and the actual amount of production should be plotted.

The relation of the actual production to the estimated production would show whether deliveries were being met or not; the relation of the average working force to the actual production would show the probabilities of the future, thus indicating the necessity for increasing or decreasing the payroll force.

The relation of the average working force to the actual production would also indicate any marked increase or decrease in efficiency and the need for further investigation in the latter case.

The chart would also show the effect of extreme weather conditions or a strike over any given period, thus substantiating a claim for extension of contract time for such delays.

It is the primary purpose of these two charts to point out just such conditions as these and the charts which follow are for the purpose of assisting in locating the reasons for unsatisfactory conditions when they occur.

The actual number of placements, terminations and transfers in and out by departments is shown in Fig. 3. The placements are shown in solid black and the number of the transfers into each department by the

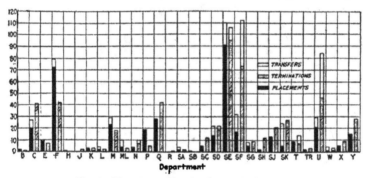

Fig. 3—Placements, Terminations and Transfers

addition of an open block. The number of terminations is shown cross-hatched and the number of transfers out is shown by an open block in similar manner to the transfers in.

The reasons for terminations are shown on the chart in Fig. 4, in the order of their relative importance.

The chart in Fig. 5 shows the percentage of absenteeism by departments, and the average of the plant as a whole.

Personnel Research. The following is quoted from an article by Eugene J. Benge in *Industrial Management*, May, 1920.

The maintenance of the working force is, of course, one of the big problems of the personnel department and is therefore a problem which challenges the attention of research. Probably one of the greatest needs in the personnel field to-day is for adequate measures of production. This is certainly not a need that can be solved by guess, and the trial and error method is usually rather costly. Establishing proper measures of production is a question that demands the best methods that science can offer for its solution. Studies of conditions that have

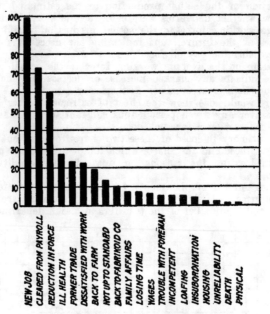

Fig. 4—Reasons for Terminations

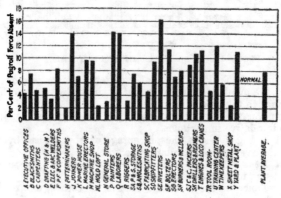

Fig. 5—Absenteeism Chart

been passively accepted, such as regular absenteeism, will often startle by their obvious significance.   The accompanying Fig. 6 indicates actual attendance conditions in a shop of one of our large corporations. It was calculated that the absenteeism limited the possible efficiency of this shop to 91 per cent.

Again, studies of accidents will usually reveal facts such as indicated in Fig. 7, thereby opening up great possibilities for the study of fatigue and accident prevention.

So far we have considered only factors which are under the control of the plant. But there are many outside conditioning factors which warrant the attention of the research director because of their possible influence. The cost of living in connection with community living standards is probably the most important, for these affect the vital question of wages. It is no simple task to arrive at accurate conclusions concerning the cost of living in any single community—the national figures published by the Department of Labor are suggestive but cannot be used directly for this purpose. A chart similar to that in Fig. 8 should prove illuminating to the management which believes in the fair deal. .

A survey of the educational and recreational opportunities of the community which supplies the plant with its workers will always afford valuable data to the concern which stands ready to utilize such information.

Analogous to studies of community conditions would be one of the sales territories covered by the company salesmen, for there are many points in common. The director of personnel research would be well equipped to carry on such an analysis for the sales department.

The research director must always be prepared to defend with cold hard facts the various conclusions and recommendations which he may make.

The procedure of the research department may readily be divided into four parts:

1. Selection of the scope of activities and subsequent definition of each problem.

2. Planning for each individual problem.

3. Execution of plans, having various phases handled by experts.

4. Interpretation of results without bias.

The use of statistical method is one of the most important aids of the personnel research director. The mind is unable to grasp the significance of larger facts or underlying principles involved in a mass of data. Statistical procedure reduces a potpourri of separate facts to a unit statement which the mind can readily grasp.

From the collection of many data there can readily be determined standards of attainment by which to value and to guide current performance. This is essential to adequate measures of production.

One of the biggest possibilities which the science of statistics offers industry is the method of measuring relationship between two seemingly disparate factors. Suppose we want to measure how well a trade test is selecting men for a certain position. We can plot the relationship between the test and actual performance as shown in Fig. 9. It is very evident that those who made high scores in the test also made high records of production, with the one exception, M who made high in the test but was only an average producer. But in order to compare the value of this test with some other test for the same purpose,

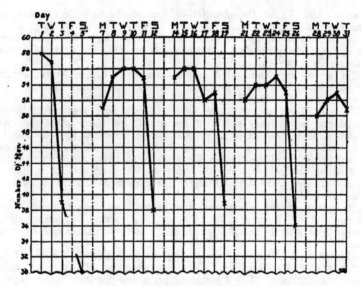

Fig. 6—Daily Attendance, Blacksmith Shop, July, 1919

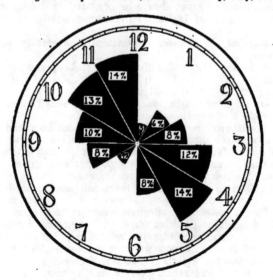

1,873 Accidents reported in one year in State of
Illinois

Fig. 7—Percentage of Accidents Each Hour of Working Day

there is needed some simple *measure* of relationship, and this is supplied by what is known as the coefficient of correlation, obtained from the application of mathematical formulas. In the group of 20 workers shown in the accompanying figure, worker *M* should be observed carefully, for there is something which prevents him from producing to his probable capacity.

Graphical presentation of facts is a great adjunct to statistical procedure, for it substitutes pictures for words. How long would it take one to present verbally the data which are shown on the accompanying charts?

Chart Records for the Employment Department. The following is from an article by Russell Waldo, published in *Machinery*, February, 1919, entitled "Use of Charts in the Employment Department":—

Fig. 10 is shown merely as an illustration, but the full set of charts for keeping records of the employment department may consist of five or more similar to the following:

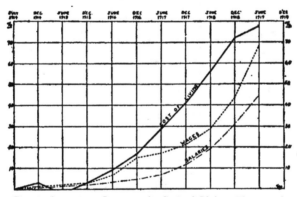

Fig. 8—Percentage Increases in Cost of Living, Wages and
Salaries in the ——— ——— Company

Chart 1 should be similar to the one shown in the accompanying illustration and gives the number of machines idle, operating, and new machines. The yellow line shows the new machines added; the red line, the number of machines idle from lack of operators; and the green line, the number of machines in operation. The total number of machines is recorded by the black line.

Chart 2 shows the status of employes. The full black line represents the total number of men on the payroll and the black dash line, the number of women. The full yellow line shows the number of men sick, and the dash yellow line, the number of women. The number of men present is represented by a full green line and the number of women present, by a green dash line, while the full red line represents the number of men absent, and the dash red line, the number of women absent.

Chart 3 should record the weekly labor conditions. The number of employes discharged should be shown in black lines, and the number leaving voluntarily, in red lines. Green lines should be used to show the number of persons hired. In all cases the full lines refer to men employes and dash lines to women employes.

Chart 4 should record the work of the employment department. It should contain a record of the number of men interviewed, which should be shown by a full yellow line. The number of men hired should be shown with a full green line, while the number of men needed should be shown in a full red line. The same color lines should be used to record the conditions in the women's employment department.

Chart 5 is of importance in that it shows the attitude of employes

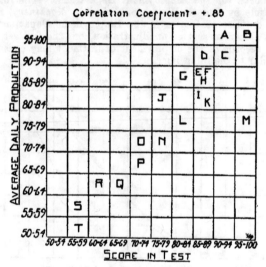

Fig. 9—Comparison of Test Scores and Production of
20 Workers

toward their work, and is more in the nature of a personal record than any of the other charts described. A record of the number of employes receiving promotion should be shown upon this chart with yellow lines. The number of employes studying for promotion should be shown with green lines, while the number of employes not studying for promotion should be shown in red. Some interesting percentage figures can be obtained by comparing the total number of employes, shown by a black line, with the number of employes studying for promotion as shown by the green line. While the set of charts described forms a complete record for the employment department, variations can, of course, be adopted to meet special requirements.

Labor Turnover. E. Goldberger, in an article, "Labor Turnover— Discussion," published in *Industrial Managment*, November, 1918, states, in part, that the Packard Motor Car Company, of which he is efficiency engineer, have investigated this proposition repeatedly and that they have adopted and applied the following method, which two charts, Figs. 11 and 12, help to make clear.

Fig. 11 shows:

A. Grand total of men employed in the division.

B. Total number of men hired during one week preceding the date marked on the horizontal co-ordinate (this includes new men, rehired, re-instated and transferred in).

C. Total number of men paid off during one week preceding date marked (includes, resigned, laid off, discharged and transferred out).

D. Total number of men hired during four weeks preceding date.

E. Total number of men paid off during four weeks preceding date.

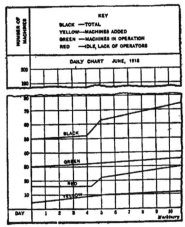

Fig. 10—Chart Used in Employment
Department

For figuring the labor turnover I have adopted a new standard that gives a better index as to the labor condition in a division.

1. Instead of one week's figure for hired or paid men, I take the total of four consecutive weeks, which then gives an evener slope in the curves expressing turnover tendencies.

One week is too short a period from which to draw any intelligent conclusion, since the variations between two following weeks might represent a couple of hundred per cent.

2. The total number of men that have been hired and been paid off, *i. e.*, actually replaced, represents a complete turnover, and if there be an excess of men hired over paid off, they actually are to be accounted to increase in force, while if there be an excess of men paid off, they represent a decrease in force.

The term "fluctuation of labor" (+ if increasing the force, — if decreasing) is used to express these figures, since it leaves the term "turnover of labor," to represent only what its name implies.

It is important to make this distinction, since avoiding a higher labor turnover calls for one type of preventatives in the employment and factory departments, while a high fluctuation is either unavoidable or the preventative means must be taken in the production and sales departments.

3. To avoid using the information after it has become one month old, from month to month I figure the turnover and fluctuation every week for the past four weeks; this gives 52 chances for intervention instead of only 12 per year.

The labor turnover, then, is the ratio between the total *replacements* in a division during the four weeks preceding the date of figuring, and

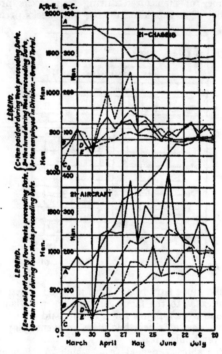

Fig. 11—Chart of the Vital Statistics of
Employment

the grand total of men employed (on our books) at the end of that period. This figure multiplied by 13 gives the labor turnover percentage reduced to a yearly basis.

The above rule naturally implies that if a man has left for certain reasons, and another one has been hired during the four weeks following his departure, a *replacement* has taken place which has to be accounted into turnover, but if he be not replaced at all, or replaced only after more than four weeks, it is accounted as fluctuation.

Fig. 12 gives both the turnover and the fluctuation (above the 0 line if increasing, below if decreasing the department's force). They are both figured in per cent. (*i. e.*, taken over the total number of men employed in that division during that week). The figures on the left side of the column read the percentage for the elapsed four weeks (almost a month) and those on the right side the same, reduced to a yearly basis.

The prevailing practice in figuring labor turnover is: Multiply the number of men hired (or leaving) each week by 52 and divide the result by the grand total of employees. This leads to wrong conclusions since it

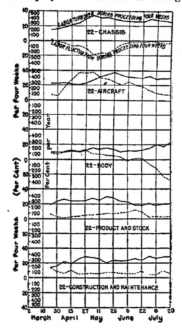

Fig. 12—Chart of Labor Turnover and Labor Fluctuations—Two Departments

is neither the hired, nor the leaving men that constitute the basis of labor turnover.

Both methods (hired or leaving men as basis for turnover) and their error appear in Fig. 11, where in two different divisions the conditions are directly opposed. Either way of figuring would affect, considerably, the factor representing turnover. Actually, the true turnover cannot be so different in two divisions under the same management and same scale of wages.

Of course, the same conclusions could be drawn if, instead of two divisions, we would consider two plants.

Men Hired vs. Men Reporting. Earl S. Morgan in *Industrial Management*, May, 1920, wrote an illuminating article about workmen who accept jobs and then never go to work. He says in part: I am not so certain that every personnel manager realizes the steady "leak" these "failures to report" are causing in our hiring costs. We are all "sold" by this time on the cost of labor turnover of applicants employed. Have you ever considered how much loss *quitting before they start* causes?

Let us see what may be involved:

1. All the time and effort securing, interviewing, and selecting the applicant.

2. Introduction to the foreman and the job.

3. Medical consultation and, perhaps, physical examination.

4. Arrangement for personal accommodations, such as locker, etc.

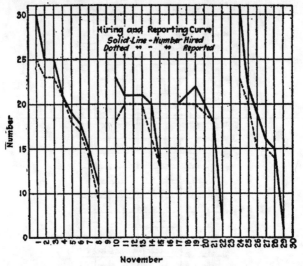

Fig. 13—Curves Showing Number of Applicants Reporting

5. Machinery is set in motion to place the prospective employee on the payroll.

6. Plans are made to utilize the employee's services and when he fails to report such production is delayed; very often a part of the equipment is forced to remain idle and schedules are disrupted.

7. Often other acceptable applicants are dismissed and a new round-up of material has to be started.

Now if you will estimate for your own conditions the investment in a prospective employee which becomes a complete loss when he fails to report, I think you will agree with me that the "leak" is worth attention.

As a first step in attacking the problem, I would suggest that a running chart be kept showing the condition daily. See Fig. 13. This will visualize the situation and show the progress that is being made in solving the difficulty.

Mr. Morgan then analyzes the probable causes of "failures to report" and recommends the "reinterview" as a remedy.

The Graphic Chart used in an Effort to Improve Punctuality. The following notes are taken from an article by George O. Swartz in *Industrial Management*, March 1, 1921: It will be sufficient for the purpose of this article to show the effects of the appeal of self-respect and approbation when applied to plant punctuality. Fig 14 shows the results of our special efforts in this respect from March 1 to November 1, 1919. The shop people came in the morning at seven o'clock, and the office workers at eight o'clock, but during the week ending March 1, the office workers reported punctually only 64.08 per cent. of the possible number of times, while the shop workers reported punctually 90.64 per cent. of the time. There were two reasons for this: First, the shop workers were docked for time lost, or if piece workers, failed to earn their maximum amounts, but the office workers suffered no reduction in their earnings whether late or not. Second, some of the office heads were men in such high positions, indeed executive positions, that they did not observe rigidly the office schedule, and clerks in their departments were concerned chiefly in getting in just ahead of

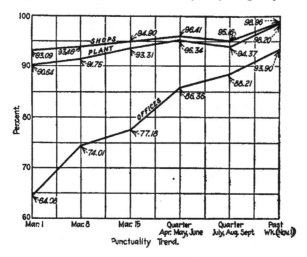

Fig. 14—Effect of Efforts to Improve Punctuality

the "boss." But it will be noted that, after all, rapid improvement was made, the punctuality of the plant rising from 90.64 per cent. on March 1 to 98.20 per cent. on |November 1. A slight dip will be noticed in the plant average during July, August and September. This is probably due largely to the fact that our foremen were taking vacations in turns within this period, and the supervision was not quite so close.

Special Personnel Department bulletin boards were posted conspicuously throughout the plant and offices, one in each department. So far as possible, these bulletin boards were located over drinking fountains. They were visited each day, and any notices posted by

other than the Personnel Department removed, and any notices of our
own that may have been removed were replaced. No bulletin was
permitted to remain on the board after its purpose had expired.
We furthermore varied the general appearance of the notices by
changing the color of paper, size, shape and arrangement, so that a
casual glance would disclose a new notice. Sometimes we would

```
┌─────────────────────────────────────────────────────────┐
│           P U N C T U A L I T Y   R E C O R D            │
│              Week Ending December 6, 1919                │
│                                                          │
│  Shop or Department                      Times having    │
│                                          100% since      │
│                                          Sept. 12,1919   │
│  ------------------------------------------------------  │
│                                                          │
│  Watchmen  -   -   -   -   -   -   -   100% -  15         │
│    Machine #1   -   -   -   -   -   -   100% -   1        │
│  Motor Truck  -   -   -   -   -   -   100% -   3         │
│  Bolt  -   -   -   -   -   -   -   -   100% -   1         │
│  Inspectors  -   -   -   -   -   -   100% -   8         │
│  Upholstery  -   -   -   -   -   -   100% -   7         │
│  Forge  -   -   -   -   -   -   -   99.63% -  0          │
│  East Yard - Lumber - Steel Stores -  98.93% -  1        │
│  Die  -   -   -   -   -   -   -   98.63% -   3          │
│  Steel Car & Prep.  -   -   -   -   98.51% -  0          │
│  Spring  -   -   -   -   -   -   98.44% -   1          │
│  Tool Room  -   -   -   -   -   -   98.30% -  1          │
│  Sheet and Press Metal  -   -   -   98.63% -  0          │
│  Brass  -   -   -   -   -   -   98.10% -   3          │
│  Shipping  -   -   -   -   -   -   98.00% -   4          │
│  Erecting  -   -   -   -   -   -   97.96% -  0          │
│  Power  -   -   -   -   -   -   97.95% -   3          │
│  West Yard  -   -   -   -   -   -   97.67% -  0          │
│  Finishing  -   -   -   -   -   -   97.59% -  0          │
│                                                          │
│  General Average  -   -   -   -   -   97.12%             │
│                                                          │
│  Seat Upholstery & Seat Metal  -   -  97.09% -  1        │
│  Repair & Maintenance  -   -   -   96.85% - 0          │
│  Paint & Varnish  -   -   -   -   96.65% -  0          │
│  Truck  -   -   -   -   -   -   96.52% -   0          │
│  Wood Mill  -   -   -   -   -   96.28% -  0          │
│  Cabinet  -   -   -   -   -   95.98% -  0          │
│  Drawing Room  -   -   -   -   94.60% -  0          │
│  Store Room  -   -   -   -   -   93.63% -  0          │
│  Equipment  -   -   -   -   -   93.33% - 0          │
│  Electrical  -   -   -   -   -   88.27% -  0          │
│  Pattern  -   -   -   -   -   87.14% -  0          │
│                                                          │
│                   ••••••••                               │
│  The boss won't be a crank if you are a self-starter     │
└─────────────────────────────────────────────────────────┘
```

Fig. 15—Record of Punctuality by Departments

post a cartoon or a verse or a bit of shop news. The thing of prime importance was to get the men to read the notices—and they did.

The initial step in our punctuality campaign was publicity of departmental records. We sent around a note announcing that such information would be posted, and would be interesting because of the wide
variations. Nothing else was said, but the punctuality next week rose
all along the line—an improvement of from 64 per cent. to 74 per cent.
in the offices, and a small gain in the shops, which were already doing
well. Fig. 15 shows the nature of these bulletins. The arrangement is
from 100 per cent. downward, in the order of the declining excellency.

We found that by posting our bulletins at a definite time each
day the men became accustomed to look for the bulletins at that special
time. The noon hour was especially favorable to our purpose, for then

the men could gather around the boards and take time for discussion of the relative merits of the departments.

An Application of the Graphic Chart to Cutting Training Costs is described in an article by J. R. Sedgwick in *Factory*, January 1, 1921, as follows:

In most factories the policy is to hold the learner in the training class until he "makes out" on piece-work. This encourages the nursing along of men unadaptable to the work, and the spending of considerable time, money, and patience in attempting to make piece-workers out of them before they are either transferred to some other machine requiring less skill, or given up as hopeless, and honorably discharged. In most cases such men get disgusted with the job and quit.

In some cases it is very evident from the first that the employee hasn't the proper coördination between mind and muscle to make an average piece-worker, but because the new employee is trying to do his best, and also because the foreman is human, the man is allowed to keep on trying. Sometimes such learners become piece-workers, but such cases are rare exceptions. The money which is wasted on the men who do not "make out" does not pay for those who become piece-workers from this class.

To find whether or not there is a possibility of predetermining the later efficiency of the worker by his production during the first few days, a careful analysis of individual operations was made.

The records of all men started on the job were studied. They were then separated into two groups; namely, "those who become piece-workers," and "those who left during the training period." Each group was taken separately, and a composite chart, Fig. 16, was made showing the class progress.

These studies proved that there are some jobs on which it is possible to predict the ability of the new worker. These consist of operations requiring a continuation of learning throughout the training period.

Such operations are usually complex in motion and require the remembering of many elements of specifications.

There are some operations which show no differentiation between the men who quit and the men who stay. Such operations, therefore, do not permit any prediction. In these, the knowledge element is very small and the "training period" is mainly an acclimating period to get the employees accustomed to the working conditions of the job.

Luckily, the more important operations usually fall into the first class, and this permits the development of progress charts for them. In the operations which come in this class there are five graduations of learners, as follows:

1. Exceptional men
2. Safe men
3. Average men
4. Subnormal men
5. Unadaptable men.

1. The exceptional men are learners who during the first few days are much above the average. Their main desire is to get the money, and get it quick. Such men have the ability to learn very rapidly, but when

the job is learned they lose interest. They are usually spasmodic in their work, and are very easily discouraged, especially if they should happen to have an "off day." As soon as it is seen that an employee falls into this class, the instructor's main function is to act as a governor on the speed of the learner, and to see that his production doesn't go "sky-rocketing" one day, and "tobogganing" the next. These exceptional men are valuable to an industry, and should, whenever possible, be transferred to development work, preferably on hourly pay or salary.

Unless special attention and coaching is given such learners by the instructor, they very seldom develop into desirable piece-workers.

2. Between the "exceptional men" and the "average men" there is a class which can be called "safe." They are learners above the

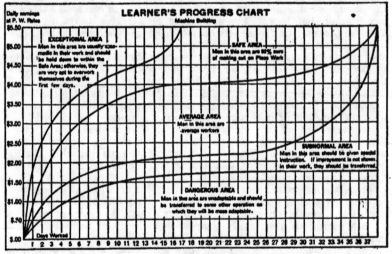

Fig. 16—Learner's Progress Chart

average, yet they do not fall into the "exceptional" area. Men in this class are reliable, and do not need the attention required by those in the other four classes.

3. Average men are as the name implies. The large majority of workers fall into this class.

4. These learners require special attention from the instructor. They might be called "doubtful," for the chances of even making out on piece-work are considerably against them. There are enough of this class who finally make out to keep them from being classed as hopeless; however, if a learner falls into this class over four days in succession, he should be transferred to some other operation on which he is more adaptable. By so doing, a great saving of time and money can be made.

5. Men in this class can be called "hopeless" provided they are in this

group more than three days in succession. Such cases should be immediately transferred to some other operation requiring the type of skill for which they are adaptable. Experience proves it to be uneconomical to try to make piece-workers of these men.

The practical application of the plan works exceptionally well. When an operation is analyzed, a chart is made, showing the different areas of the above classes, Fig. 16. A copy of this chart is given to the in-

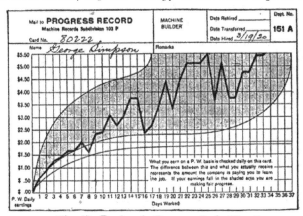

Fig. 17—Employee's Progress Card (Front)

Fig. 17—Employee's Progress Card (Back)

structor on the job. With this chart he knows on which men to concentrate his efforts. There are definite limits between which a worker may be and by playing the game according to rule the instructor has a good shot of rightly judging his man.

By strictly following instructions on the chart, the idea proved so successful that it was furthered by putting such a chart, Fig. 17,

on each learner's machine. The chart was printed on individual progress record cards. This card shows the "safe" and "average" areas
as shaded, and the employee is given to understand that as long as his
production stays in the shaded area he is making fair progress.

When the new employee arrives on the job, he finds a progress
card, hanging in a rack on the machine on which he is to work. Each
day, at the end of the shift, the instructor marks the new employee's
production on the card, so when he arrives the next day he has a
visual record of his previous day's work. It has been found that this
gives a big incentive for the man to beat his previous day's work.

This chart, of course, gives the same information that the progress
card does. The plotted curve enables the foreman and the student
to more readily visualize the man's progress. And the fact that the
beginner's curve is plotted between lines which show whether or not
his work is above the average emphasizes his deficiencies more than a
mere comparison of figures.

This incentive has so shortened the learning period on some operations
that it is necessary to revise the charts periodically. This shortening
of the training period, of course, means less money spent on training,
less money lost on labor turnover among new employees, greater production, and a better satisfied working force.

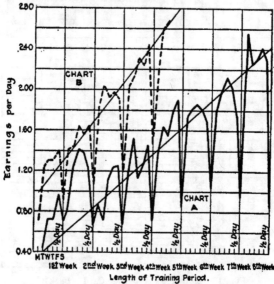

Fig. 18.—Chart A Shows the Progress of New Worker at Start. Chart B Shows
Progress after Instructor Was Given Period of Training.

The Use of the Graphic Chart in Factory Training work is shown
in an article by James F. Johnson in *Industrial Management*, February,
1920, part of which follows: The accompanying charts have been
compiled from data showing results obtained in training new workers
in several different branches, according to principles here referred to.

The length of time usually taken to show proficiency in each of these branches was from 12 to 15 weeks under the old plan of "breaking in." Under an organized instruction plan this was reduced to 5 and 6 weeks.

Chart A of Fig. 18 shows how a new worker progressed while in charge of an instructor very capable as a tradesman but undeveloped as a teacher. The irregularity of the progress from day to day indicates immediately what might be expected under such conditions.

However, after a careful analysis of the work was made and the instruction material (which, by the way, was the same as before), was rearranged according to its learning difficulties, this instructor was given a period of training in how to best impart the necessary trade knowledge. The results he then obtained were surprisingly good and the strain on both the learners and the instructor was much less than before.

Chart B of Fig. 18, shows results obtained after this change. The steady rise in this curve shows good progress and an early attainment of highest rate on the chart. The chart of Fig. 19 shows an inter-

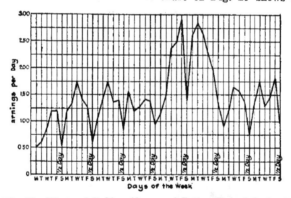

Fig. 19—Effects of Sudden Absence of Better Trained Instructor

esting condition where the learner after plugging along for a few weeks without much method was suddenly given attention by her instructor who rearranged the order of her work and taught her according to instructions he had already received in teaching methods. The change in accomplishment for the period is very interesting as well as conclusive.

It so happened, however, that it became necessary for this instructor to be away from his job and another instructor, untrained, took his place. Almost immediately the new condition was reflected in the drop of the curve to its former pace.

Employees' Benefits. Charts are a great help in explaining the methods and features of various employees' benefits, such as insurance, pensions, loan funds, etc. A typical example is Fig. 20, which, with the following explanation, is taken from *The Automobile*, April 5, 1917.

There are other benefits not mentioned here, this being taken merely to show the figure.

Joint payment by manufacturer and worker is the feature of the Fell plan, which provides for old age and other times of need. The joint-payment factor makes continuance of relations between the manufacturer and the worker essential to the working out of the plan, so that the system is in essence turnover insurance with the granting of pensions, death, accident, and sickness features as a by-product.

Preliminary to the paying in of any funds, an association is formed in the factory composed entirely of employees. The membership is divided into departments, each department selecting its own local secretary. The local secretaries select the board of directors for the insuring system, and the directors appoint such standing committees as may be required. This type of organization is adopted to put the control of the plan in the hands of the employees.

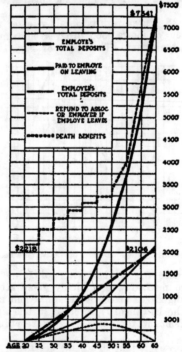

Fig. 20—Showing Workings of Fell Plan
When Employee Enters at 20, Wages
$18

Each worker pays a flat rate of 5 per cent. of his wages into the insurance fund. The employer contributes to the same fund on a graduated scale which hinges upon the length of time the employee has been with the company. The man who is a new employee pays a larger amount than the manufacturer, but the converse becomes the case as the term of service lengthens. The worker is paid for his investment by the various insurance privileges to be described, and the manufacturer by the saving of waste in changing the personnel of the force.

The men are divided into 5-year groups. The employer pays into the fund 1 per cent. of the wages of those who have been with the company less than 5 years, 2 per cent. in the case of those who have been employed from 5 to 10 years, and so on. The following table shows the schedule:

Years' service of employee:	1–5	5	10	15	20	25	30	35	40
Percentage of employee's wages paid into insurance fund by the company	1%	2%	3%	4%	5%	6%	7%	8%	9%

Inasmuch as the employer has no actual cash claims on the fund and derives therefore no direct financial benefit, his payments from the first year represent 20 per cent. interest on the total invested, and constantly increase. A large number of employee's privileges can therefore be paid for under this system.

The easiest way to see how the plan works is to take a typical case. A young man starts work with a company when he is 20 years old. He gets $18 a week, and retires at 65 years of age. Without figuring that the young man might get a raise during 45 years of service, his financial situation from time to time is as follows:

He will have paid in $234 at the end of the fifth year. For this he gets: 1—A life insurance policy for $2,217 during the period. 2—. The right to borrow $168 in case of sickness at the end of the fifth year. 3—A surrender value of his contract of $126.

At the end of 20 years the privileges are proportionately much greater, due to the compound interest which the money may have earned and the graduated scale of payments for the employer. The young man at the age of 40 then has paid in $936.

He is carrying a life insurance policy worth $2,977, can borrow in case of sickness $1,268, or surrender the contract for $1,008 cash.

When this employee has reached the age of 65 he can retire from active service with a pension of $677 a year for the rest of his life, and if he should die before he had received the pension for 10 years it will nevertheless be paid until a 10-year period has elapsed to his estate. In lieu of a pension he may have a paid-up insurance policy of $7,341, and may withdraw this sum if he so desires.

CHAPTER XXXI

GRAPHIC CHARTS FOR PREDICTION AND TREND

Before reading this chapter, the section in the latter part of Chapter XI, entitled Forecasting by Means of the Ratio Chart, should be reviewed.

The abstracts from articles which follow have been presented because they suggest ideas for the use of charts in connection with prediction, rather than because the charts illustrating the articles are the best ones to use. In fact, the authors believe that most of the charts shown here on plain ruling should be on ratio ruling, but they have been left unaltered in order to preserve the work of the original article.

It is believed by many that, if, instead of plotting actual quantities, the quantities are reduced to a percentage basis, and the percentages plotted, the plain ruling is quite all right, because then one may read the percentages from the scale on the chart and so not need to estimate or measure them as would be done if the quantities themselves were plotted on ratio ruling.

There are at least three good arguments against this: 1. There is a very considerable amount of labor required to reduce the quantities to a percentage basis, which is not necessary when ratio ruling is used. 2. When the quantities are reduced to a percentage basis, and the percentages plotted, the chart does not show the actual quantities, and it is often very desirable to have them. 3. While it is possible to read off the percentages directly, on plain ruling, the picture is still a misleading one for the reason that the same numerical difference in percentage is always represented by the same vertical distance on the chart. In other words, a rise from 10 per cent. to 20 per cent. would look just the same on the plain ruling as a rise from 80 per cent. to 90 per cent. The numercial difference is the same in both cases—10 per cent., but the percentage difference is 100 per cent. in the first case and 12½ per cent. in the second. Therefore, percentages should be plotted on ratio ruling to obtain a true picture of the relative differences in per cent.

Planning. The use of the graphic chart in planning ahead is shown in an article by J. W. Scoville in *Factory*, July, 1921, in part as follows:

Forecasting is a fascinating game. It is distinguished from pure guesswork in that it rests upon extensive analysis. At the end of each month, we draw up an estimated future balance sheet for each of the four following months. These forecasts agree remarkably with the actual occurrences when each forecasted month comes around, as shown by the chart comparing forecasts with actual conditions, Fig. 1.

Business forecasting involves three important steps. First there is the collection of data. Then comes the analysis of the past records to determine their interrelations. This work is made easier by the use of graphic charts. Finally there is the application of the fundamental principles discovered by analysis to determine the future position of the business.

In forecasting the business of the American Writing Paper Company, the form shown in Fig. 2 is filled out. At the left are 25 items to be forecasted, and each quantity is worked out for four months in advance.

The first step is to forecast sales, in dollars and in pounds of paper. Curves are available showing sales by 6-month periods for 20 years and

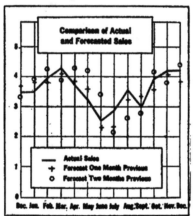

Fig. 1—Curve Showing How Forecasted
and Actual Sales Compare

by months for several years. Trade Association reports and general business conditions are studied. A conclusion is reached as to whether sales are likely to advance or to recede, and finally, sales are estimated for four months in advance. This estimate is the keystone to the arch as all subsequent estimates will be seriously in error if a mistake is made in prospective sales. Having estimated prospective sales, and knowing from market conditions whether it will be advisable to increase or decrease stocks on hand, the production can then be forecasted. The fourth step is to estimate the trend of raw material prices, as this will influence the amount of sales and the value of the inventories.

As we know how much paper we expect to make each month, and how much we expect to sell, we are in a position to compute the seventh item, the value of paper in inventory at the end of each month. Likewise, knowing how much of each raw material we have in stock and knowing the consumption required each month to keep up production, we can estimate the sixth item, the amount of raw materials to be purchased each month and the amount in inventory at the end of each month. From the financial condition of the company and the monthly purchase of raw materials, it is now possible to estimate the probable disbursements to reduce the accounts payable, item 18.

The monthly payroll, item 19, is estimated from the curve shown in Fig. 3. On this chart each dot shows the weekly payroll for past months with the corresponding production. After noting whether there are four or five pay days in the month, it is possible to estimate the disbursements for payroll.

In this manner we proceed to build up a hypothetical balance sheet for the four months to come. Certain expenditures may be more or

FORECAST FOR THE AMERICAN WRITING PAPER COMPANY

ISSUED MAY 1, 1921.

	May	June	July	August
1. Net Income from Sales	--	--	--	--
2. Pounds of Paper Sold	--	--	--	--
3. Pounds of Paper Made	--	--	--	--
4. Prices, All Materials, end of month	--	--	--	--
Current Assets	May 31	June 30	July 31	August 31
5. Total	--	--	--	--
6. Materials and Supplies	--	--	--	--
7. Paper	--	--	--	--
8. Accounts Receivable	--	--	--	--
9. Cash	--	--	--	--
Current Liabilities				
10. Total	--	--	--	--
11. Bank Loans	--	--	--	--
12. Accounts Payable	--	--	--	--
13. Excess of Current Assets over Current Liabilities	--	--	--	--
14. Ratio of Current Assets to Current Liabilities	--	--	--	--
Receipts	May	June	July	August
15. Collections	--	--	--	--
16. Total Receipts	--	--	--	--
Disbursements				
17. Total	--	--	--	--
18. Materials and Supplies	--	--	--	--
19. Pay Roll	--	--	--	--
20. General Administration and Overhead	--	--	--	--
21. Advertising and Selling	--	--	--	--
22. Interest	--	--	--	--
23. Taxes and Other Disbursements	--	--	--	--
24. Receipts over Disbursements	--	--	--	--
25. Net Profit	--	--	--	--

000 omitted

Fig. 2—Items Forecast Each Month for the Next Four Months

less erratic and dependent on the psychology of the executives in control. The more nearly the statistician is in touch with the plans of the executives, the nearer the forecasts should tally with the subsequent events. The estimates are based on several basic figures from independent sources, but after they are made up we put them to interior tests.

While the forecasted balance sheet will never be exactly fulfilled, it should be consistent. There are about a dozen consistency tests that must be fulfilled, 1. The cash on hand at the first of the month plus the collections less the disbursements equals the cash on hand at the end of the month. 2. Accounts payable at the first of the month plus purchases less disbursements to creditors equals accounts payable at the end of the month. 3. The value of paper inventory at the first of the month plus the cost of paper manufactured less the cost of paper removed from stock plus (or minus) changes in value due to market fluctuations equals value of paper inventory at the end of the month. Every possible consistency test should be formulated and applied to the figures in the forecasted balance sheet.

Graphic devices have been used to analyze certain items. In forecasting the accounts which will be receivable at the end of each future month, we use the intimate relation that exists between the accounts receivable and sales. An analysis of our accounts for a great many months shows that payment for the sales of each month is spread out mostly over the following three months, with a varying proportion in

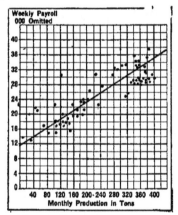

Fig. 3—Curve for Estimating
Monthly Payroll

each of these months. Similarly, the accounts receivable in any one month are made up of parts of the sales of previous months. We find that if we take two thirds the sales of the present month, half the sales of the previous month, and one-third the sales of the month before, and multiply this total by a certain percentage, this gives a base line that is equal with a fair accuracy to the accounts receivable at the end of the month. In fact the accounts receivable do not vary from this figure more than 20 per cent., and usually not more than 10 per cent. Fig. 4 shows the fluctuations in accounts receivable as compared with the figure calculated from three months sales in this way. Knowing our sales, for the current months, and having forecast them with fair accuracy for a few months in advance, we can use Fig. 4 to forecast our accounts receivable at the end of each future month, modifying the result according as we see that prosperity or dullness will make corrections.

A device which is useful in forecasting profits is shown in Fig. 5. Each dot shows the net profits and the gross sales for a certain month. These dots tend to form a straight line, or to cluster along a straight line. Placing a ruler in the best position between the dots, we draw such a line, the slant of which gives us a ratio of possible profit to

possible sales, by which we can forecast future profits from the estimated volume of sales. A horizontal line shows where the profits fall
below cost of operation and become losses.   This line corresponds to the
volume of production which we have found necessary to keep the plant
efficiently busy; for, of course, if men have to be retained while pro-

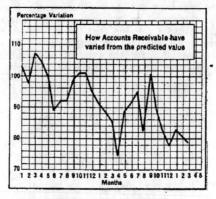

Fig. 4—Curve Showing How Actual Accounts Receivable Compared with Calculated Amounts

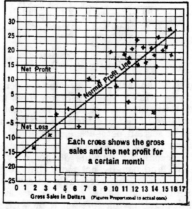

Fig. 5—Curve for Estimating Future
Net Profits

duction is not up to capacity, the overhead expense still continues and
is not offset by sales profits.

A consistency test for profits is that the change in item 13 (excess
of current assets over current liabilities) will be the profit (or loss) for
the period, except for changes that may have occurred in the value of
permanent assets.

The forecast should always be compared with the actual results when they are known. This enables the forecaster to form a correct estimate of his batting average and it enables him to improve his technique.

The percentage of accuracy that can be attained in forecasting varies greatly with the items. Sales should be forecasted with considerable accuracy. Items like profits, which are the difference between income from sales and costs, are likely to fluctuate considerably from the estimate.

The value to the executive of a scientific forecast is obvious. The last item in the list to be forecasted is the cash balance, but this is the principal aim of the entire proceedings. Our executives have confidence in the forecast, and realize that it is not a crazy patchwork of figures, but a coherent whole, consistent, and based upon scientific analysis.

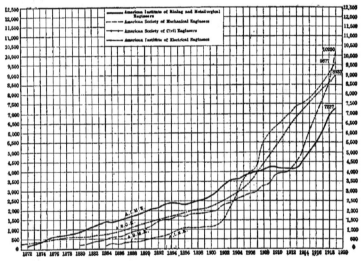

Fig. 6—Increase of Membership Curves of the Four Founder Societies

So they use this advance information of undesirable conditions and sometimes have the power to correct undesirable tendencies. Regular accounting must necessarily show only the record of past transactions. Forecasting attempts to record the transactions before they happen. In normal times, with steady prices, and a business of nearly uniform volume, the business man may use the simple rule with considerable success, that the future will duplicate the past. But the war has made such quick and violent changes, that careful forecasting has more than usual value for the business executive at the present time.

Membership Trends. Fig. 6 shows a chart taken from a report of the American Institute of Mining Engineers. This chart was given to show the trend in membership of the four important engineering societies over a period of years. It will be noted, however, that as this chart is plotted on plain ruling it does not correctly picture the trend in membership—it merely shows the numerical difference in the number of members of the four societies. From Fig. 6 it would appear that the membership of all the societies showed a marked gain during the last 20 years. It is true that the societies have gained in the number of members, but the rate of increase has decreased. This is clearly brought out by the chart in Fig. 7. Here the same data are plotted on ratio ruling. As the general trend shows a curve ascending and bending downward, convex fashion, the membership is increasing at a

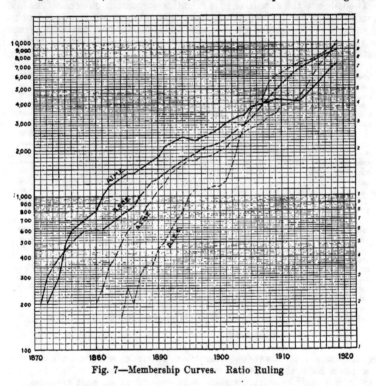

Fig. 7—Membership Curves. Ratio Ruling

decreasing rate. In other words the rate of increase is falling off. The picture presented by these data on plain ruling is misleading in that the membership appears to be increasing at an increasing rate.

Trends in Design and Manufacture. Figs. 8 and 9 are good examples of trends in design and manufacture. These curves are especially interesting to a prospective purchaser, and would also be of inestimable value to one about to start business along that line. This chart is from *Automotive Industries*, January 3, 1918, and they explain that considering all the chassis listed without regard to the output of each maker the averages deduced are averages of engineering opinion; and it is this which such analyses seek to discover. An analysis by production would throw into absurd prominence the features of a very few cars which happen to be very cheap to be made in huge amounts.

Here again, these curves should be plotted on ratio ruling to give a correct picture of the percentages of increase and decrease.

Probable Population. In *Engineering News*, November 2, 1916, Geo. H. Moore had the following :—

What the author has named the Moore expectancy curve is a graphi-

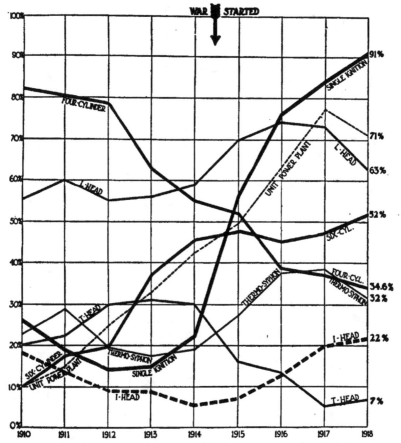

Fig. 8—Trends of the American Automobile for 8 Years, Showing Variations of Most Important Factors in Design and Manufacture

cal representation of the population a city may be expected to have after any given elapsed time in years. The curve is derived by plotting on cross-section paper the actual populations of selected cities after equal periods of elapsed years, ignoring exact dates and choosing such

values for "elapsed time in years" as may enable the respective curves
to conform most nearly to a final mean value for the group. After the
several curves have been plotted, the mean or composite curve is sketched

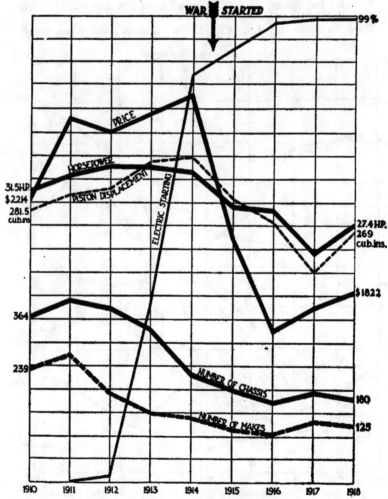

Fig. 9—Trends of the American Automobile for 8 Years, Showing Variations of
Most Important Factors in Design and Manufacture

in. This alone is traced off on transparent paper, as shown in Fig. 10
for use by itself or for superimposing upon other curves for various
comparative purposes.

To illustrate a simple use of the curve: Suppose one desires to obtain the most probable population that a city of 500,000 will possess after a lapse of 20 years. It is necessary only to mark the 500,000 point on the expectancy curve, add 20 years to the elapsed time, which in this case will give a final of 71 years, and read off 960,000.

The method can be used in almost every problem that involves either population estimate or population analysis. It can be used to predict the most probable population of a city of any size whatever at any given future time. Or, on the other hand, by comparing the plotted curve for any city with the expectancy curve it becomes evident at a glance whether the city in question is growing, or has grown, faster or slower than the average of its class.

Finally, it will be evident after a little reflection that analyses based on the expectancy-curve method are exactly as reliable as are insurance risks written on the basis of mortality tables. And since by its use the three quantities of maximum, minimum, and mean popula-

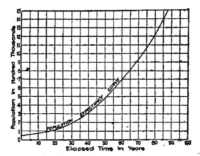

Fig. 10—Moore Expectancy Population
Curve

tions to be provided for are readily obtainable, it is claimed that the expectancy curve provides a thoroughly defensible method—possibly the first—for the exact analysis of population problems.

Predicting Yearly Earnings. From a letter, by Henry Hess to the editor of *Mechanical Engineering*, February, 1919, is taken the following:—

The executive, whether he deal with materials or dollars, is not satisfied with a mere record of what has occurred in the past, but he wants to use the past experience as a basis for consideration or production of the future. It is possible to extend a curve of monthly fluctuations to indicate a future probability, but this is neither safe nor simple.

Plotting the data as in Fig. 11 answers every question as to the past and also readily predicts the probabilities of the future.

To deal with something concrete, let us say that it is the profits of the business that are being examined, that these are being ascertained monthly, that the corresponding result for the year is desired and that the prediction is to be brought up to date or corrected each month.

In Fig. 11 each month's earnings are plotted to a curve, the lower one shown.

Over each month there is plotted the total profit to date. For instance, January shows $250, February, $300, and the total therefore to date is $550. These are shown by the upper solid curve.

To predict the probable profit for the year it is only necessary to draw a line from zero through the total earnings to date and to continue that line to the twelfth month; for example, the total profit by the end of the fourth month was $1150; the projection shows $3450 as the probable rate of profit for the year resulting from the first four months. An examination of the diagram shows that there is an improvement in

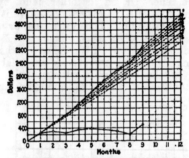

Fig. 11—Diagram for Predicting Yearly Earnings from Earnings for Part of Year

the second and third months over the first month and an improvement in rate by the end of the fourth month over the previous months. By the eighth month, however, the rate had fallen off from that of the fifth, sixth and seventh months, as is clear from the production line for the eighth month falling below that of these previous three months. The ninth month, however, showed clearly an improvement in the rate to date.

If the earnings for each month were plotted on ratio ruling the slope of the curve connecting the points would be an accurate measure of the trend and, if desired, the actual rates of increase or decrease could be read off directly from month to month.   (See Percentage Scale, Chapter XI.)

CHAPTER XXXII

Graphic Charts in the Purchasing Department

Among the necessary qualifications of an efficient purchasing agent are a thorough knowledge of prices and of demand. Price and demand are usually pretty closely related and it is often possible to foresee a probable rise or fall in price through a knowledge of demand.

Charts are a very desirable adjunct to the other records of the purchasing department. Curves showing the price fluctuations and production of the raw materials entering into the manufacture of a commodity are extremely valuable. A study of these variations will soon enable the purchasing agent to note the tendencies and to govern his buying accordingly. Figs. 1, 2, 3, and 4 are typical charts of this kind. Figs. 1 and 2 are reproduced from a bulletin of the U. S. Bureau of Labor Statistics. The prices have been reduced to a relative basis, taking the year 1913 as 100. Figs. 3 and 4 are from *Management Engineering Magazine*. The data for them were issued by the Department of Commerce.

It will be noted that these charts are all plotted on ratio ruling, as they should be, even though the figures have been reduced to a percentage basis. In this connection the following short abstract from an excellent article by Percy A. Bivins in *Industrial Management*, August, 1921, is interesting.

Purchasing agents often maintain charts of commodity prices compared with the market prices of the basic raw material entering the commodity. In Fig. 5 is a comparison of a fabric with raw cotton drawn on the arithmetic scale. Here it is noted as a general observation, that fluctuations in Cotton did not produce equal reactions in Sheeting. In the latter half of 1916 (in the part marked A-A), Cotton appeared to be soaring rapidly while Sheeting seemed to be advancing at a lower rate. The interpretation was that the sheeting market was in a favorable position. A little mental calculation showed that Cotton had increased from 13c. in July to 20c. in November or about 1.5 times and that Sheeting had gone up from 6.7c. to 9.2c or 1.4 times. The situation was therefore favorable but not as much so as had been assumed. The chart was then redrawn on a ratio basis (Fig. 6) and the relative trends over any period were exactly shown by the slopes of the curves, without the necessity of calculations, and the relative percentage changes from month to month were indicated by the vertical rises or drops. The entire series of this type of charts was then swung over to the ratio method. Other charts, as of comparisons of market, purchase and inventory prices where only a visualization of the magnitudes was desired, were continued on the arithmetic basis.

There Are Many Other Data of a more general nature which have a bearing upon prices and demand, and the well informed purchasing

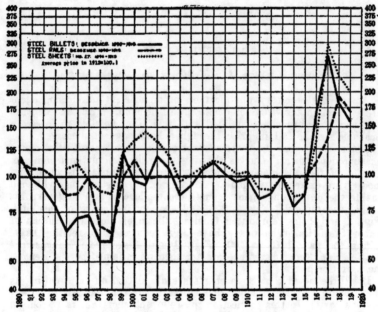

Fig. 1—Relative Prices of Steel by Years from 1890 to 1919

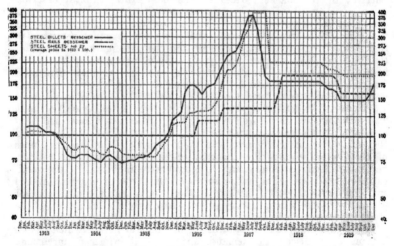

Fig. 2—Relative Prices of Steel by Months from 1913 to 1919

agent sees to it that he keeps up to date on these things. Figs. 7 and 8 illustrate charts of this character. Fig. 7 is an excellent barometer of general business conditions. When the index number of idle freight cars is large compared with average conditions it means that the in-

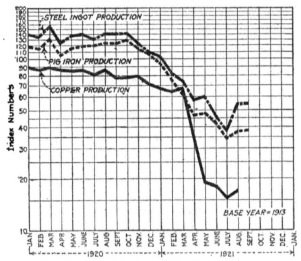

Fig. 3—Production of Steel, Pig Iron and Copper

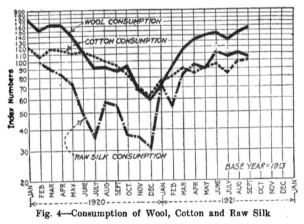

Fig. 4—Consumption of Wool, Cotton and Raw Silk

dustries of the country are inactive, and in proportion as they become active the curve of idle cars will drop.

Fig. 8 shows the curve of all wholesale prices since 1915 and indicates the direction of their trend. This chart is made up from data published

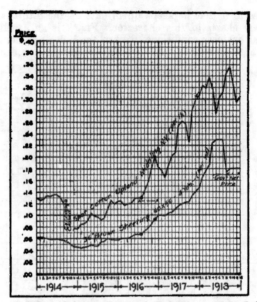

Fig. 5—Prices of Cotton, Raw and Finished, on Plain Ruling

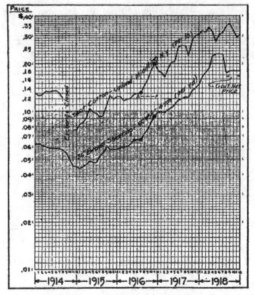

Fig. 6—Prices of Cotton, Raw and Finished, on Ratio Ruling

by the U. S. Bureau of Labor Statistics. Fig. 9 shows the various groups of commodities and the index numbers for each group.

A chart similar to that described in Chapter XXVII is very useful

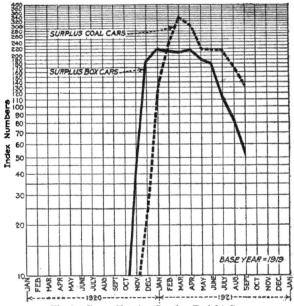

Fig. 7—Curve Showing Surplus Freight Cars

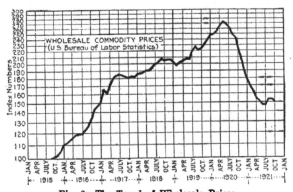

Fig. 8—The Trend of Wholesale Prices

for the purchasing department in keeping track of the material used and giving warning when it is time to order new supplies.

To Prevent Overbuying. The following, from an article, ''Nutshell

Reports for the Boss," by James H. McCullough, published in *System, the Magazine of Business*, May, 1918, illustrates the value of a graphic chart to prevent overbuying.

Fig. 10 was designed by one factory manager to help him control better the purchase and use of a certain class of materials. The heavy straight line at $3,250 shows the ideal inventory—the amount with which an excellent rate of turnover could be secured. This quantity is just about enough to run the department for a month and a half. The heavy upper line shows the real inventory. The solid lower line shows the month's purchases, and the dotted line the material used.

Owing mainly to uncertain market conditions, the actual inventory has been consistently well above the "ideal." Efforts to force it downward are evident, however, in the latter part of 1915 and throughout 1916, and again in the latter part of 1917. In March, April, May, June, and July of 1917, purchases were unusually heavy, and since the

TABLE 4. INDEX NUMBERS OF WHOLESALE PRICES, BY GROUPS OF COMMODITIES, FOR SEPTEMBER AND OCTOBER, 1921

(Compiled by the United States Bureau of Labor Statistics
Base Year 1913 = 100)

COMMODITY GROUP	1921		1920
	September	October	October
Farm products................	122	119	182
Food, etc....................	146	142	204
Cloths and clothing..........	187	190	257
Fuel and lighting............	178	182	282
Metals and metal products.....	120	121	184
Building materials...........	193	192	313
Chemicals and drugs..........	162	162	216
House-furnishing goods........	223	218	371
Miscellaneous................	146	145	229
All commodities..............	152	150	225

Fig. 9—Index Numbers of Groups of Commodities

amount used did not keep pace, the actual inventory line ascended rapidly.

The chart clearly shows that immediately after this, purchases were reduced while the amount used increased. As a result, by February of 1918 the actual inventory is as low as it has ever been.

Chart Analysis as an Aid in Buying. The following from an article "What's Behind a Good Buyer's Guess?" by A. W. Douglas, published in *System, the Magazine of Business*, May, 1916, shows the value of charts as an aid in the analysis of conditions relating to buying. See Fig. 11.

The ordering of goods is usually held to be a very simple, elemental affair largely bound up in a want book. But in truth it is one of the most complex propositions imaginable and requires, if it is to be done successfully and correctly, the utmost thought and intelligent study.

A well equipped retail hardware store in a large town may have anywhere from 10 to 15 thousand items in stock. The problem is to have as many of these as necessary on hand all the time—and no more —so that there may be no shortages when orders are to be filled, no dead stocks, and no overstocks of seasonable goods to be carried beyond the season.

The real difficulty lies in the fact that each of these items in stock, or each line of goods, is constantly affected in price and demand—or

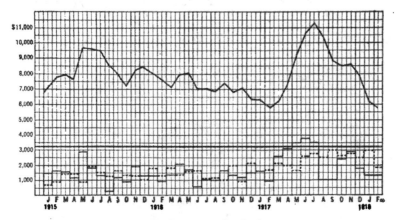

Fig. 10—To Prevent Overbuying

both—by practically everything that happens in the world. Therefore, the quantities that should be ordered or kept in stock are constantly changing.

Of all the causes which affect sales none is more puzzling than the weather. This is for the simple reason that the weather can not be foretold with any degree of accuracy more than 36 hours in advance.

I went back over the weather bureau records for all parts of the country for over 50 years, and the only conclusion I reached was that the weather runs in irregular, but none the less certain, recurring cycles, each of about 50 years. Also, that in these periods there occur the same extremes of heat and cold, of wet and dry seasons.

And it is upon these related phenomena—the reappearance of past experiences—that most of the guesses in ordering goods have to be founded. Whether you will or no, you have to do some kind of guessing as to the quantities needed. You must base your guesses on past experiences.

Here are some sales figures for typical seasonable goods:

	wet 1912	dry 1913	dry 1914	wet 1915
Rubber Hose—Feet	40,000	48,000	62,000	31,000
Grass and Grain Scythes (Dozens)	1,000	800	400	850
Ice Cream Freezers	2,000	2,100	1,900	1,400
Refrigerators	600	720	780	510

The year 1912 was characterized by abundant rainfall. 1913 and 1914 were extraordinarily hot and dry. 1915 was the wettest year known for a long time.

The figures cover the first six months of each year—the period when these goods sell heavily. Rubber hose is, of course, mostly used in dry weather, for sprinkling lawns. So the sales in that line are just what might have been expected from the prevailing weather in each year. See Fig. 11.

Both grass and grain scythes are used principally in wet weather. The comparatively large sales in 1913 are due to a curious phenomenon which illustrates the necessity of detailed study of each year. The spring and early summer of 1913 had the usual amount of rainfall.

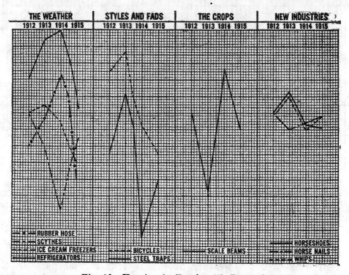

Fig. 11—Keeping in Touch with Demand

The drought did not commence until early in July when the scythe season was over. Conversely, in 1914 the dry spell commenced about the first of May and lasted all through the scythe-selling season.

These conditions likewise influenced the sales of ice cream freezers and refrigerators. They are hot, dry weather goods, and sell best in years like 1913 and 1914, and slowest in wet, cool years like 1915.

You can easily and logically connect the sale of seasonable goods with the prevailing weather conditions. You can make an intelligent study of past cause and effect as influencing your present action under similar conditions. You are no longer relying on your memory, which is usually short and treacherous in such matters.

If, for instance, one year an abnormally wet season during grain harvesting resulted in unusually large sales of grain scythes, the graphical presentation of that fact on a chart, as in Fig. 11 should prevent your orders for the following year from blindly counting on the same conditions.

Of course good or bad times are always a constant factor one way or the other. This factor in bad times tends to decrease sales, and in good times to increase sales beyond the effect of weather conditions.

Another factor which must be constantly watched by the forehanded is that of the fashions in some goods. Steel traps for catching fur-bearing animals sell best when the price of skins is high. The price of skins is in turn set by the demand for finished furs. This is in turn set by the fashions.

The years of 1912 and 1913 were years of high-priced skins and much wearing of furs. But when the summer of 1914 came, the forehanded buyer found out from some fur dealer, or from the Bureau of Foreign and Domestic Commerce at Washington, that our exports of furs to Europe were about 50 per cent. of our total production. He therefore knew that the demand for furs from this county would be cut in two. It was then obvious enough to him that the price of skins would decline accordingly, that trapping would be unprofitable, and that the sale of steel traps would fall off. And that is just what happened.

Another important source of demand variations is the wide divergence in the yield of crops. This results in a corresponding divergence in the demand for agricultural implements of all kinds.

Information regarding the promise of the crops and estimates of the yields can easily be secured from the free bulletins of the Department of Agriculture at Washington.

In all things it is necessary to check general prevailing impressions by facts and common sense, lest you go off on a tangent.

Now observation, study and thought of this sort have two purposes: In the beginning they add much pleasure and interest to a part of business usually regarded as dry and grinding in its monotony. Their second and more practical purpose is to enable you intelligently to order and handle your stock of merchandise so that you will be able to fill your orders promptly, avoid unnecessary and costly overstocks, turn over your stock often, anticipate both the coming demands for new goods and the dying out of old demands, and have seasonable goods when you need them and be entirely out of them when the selling seasons end. And the net result is that the merchandising of your stocks will become a living and profitable part of your business.

CHAPTER XXXIII

GRAPHIC CHARTS IN THE SALES DEPARTMENT

One of the broadest fields for the use of graphic charts is the sales department. There it is absolutely necessary to keep in constant touch with the progress of events. The sales manager must know from month to month, week to week, and day to day, how his sales are progressing, what the expenses of his department are, and whether his salesmen are performing efficiently.

Graphic charts are probably used at the present time more extensively in sales work than in any other branch of business. This is because they give the sales manager the information he must have, more concisely, quickly and clearly than any thing else would. They present the best means for comparison and analysis—two things which the sales manager is constantly concerned with.

For example, graphic charts will show better than any other method such important sales facts as:

(1) Total sales compared with sales "quota" or estimated sales. On the same chart may be shown the monthly variations and the cumulative total.

(2) A comparison of sales by districts—cities, states, etc.

(3) A comparison of sales resulting from various sources or methods, such as advertising, mail order, salesmen, dealers, etc.

(4) A comparison of the sales of different articles manufactured.

(5) A subdivision of sales expense—salaries, salesmen's expenses, etc.

(6) The performance of the various salesmen, such as:

Number of prospects reported.
Number of calls made.
Number of orders obtained.
Average cost per order.
Ratio of calls to prospects.
Ratio of orders to calls.

Fig. 1 illustrates (1) above. We will suppose, for example, that it had been planned to sell $6,000 worth of a certain article during the year. The average monthly quota would then be $500. In Fig. 1, the straight horizontal line represents the monthly quota of $500. The diagonal line, from 0 to $6,000, represents the cumulative quota. These two lines are drawn on the chart at the beginning of the year and represent a standard to be attained. Then, from month to month, as the actual sales are reported, they are plotted upon the chart (the dotted line) and a picture is thus obtained which shows whether the sales, both from month to month, and total, equal the mark set. The

200

plain ruling is preferable for a chart of this kind as the interest centers upon a comparison of amounts and is not concerned with percentage variations.

Fig. 2 is a graphic chart of the gross business secured by a salesman

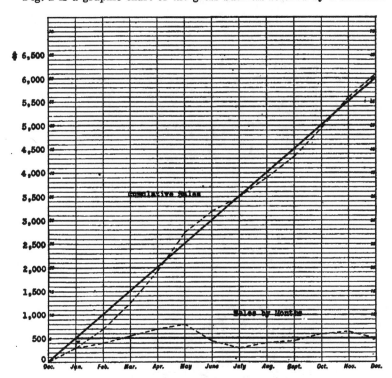

Fig. 1—Chart Showing Comparison of Actual Sales with Sales Quota

over a period of a year. This chart clearly shows the effect of the season of the year on the amount of sales. Special causes of rises and falls in the gross, such as this, may be noted on the chart to aid in the analysis of the facts.

A chart showing the relation of the number of orders to the number of calls made by a salesman is given in Fig. 3. This is an efficiency chart of the salesman. As time goes on it is reasonable to expect that the ratio of orders to number of calls will increase. If it falls it may be the fault of the salesman or of general business conditions. This may be determined by comparing his chart with those of other salesmen. The charts shown in Figs. 2 and 3 should be plotted on ratio ruling, otherwise they might be misleading and a wrong conclusion be drawn from them. The trend is what counts and ratio ruled graphic chart sheets are the only ones to correctly show it.

The following illustration of the use of the ratio chart in plotting
sales data is from an article by Percy A. Bivins in the July 1, 1921,
issue of *Industrial Management.*

In Fig. 4 comparative volumes are shown of sales of three products
having the same general ups and downs. While it is obvious from
the position on the chart that the Product A is about 3 times as large
as B and over 20 times as large as C, yet it is desired to know their rel-

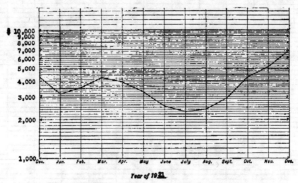

Fig. 2—Gross Business Secured by a Salesman

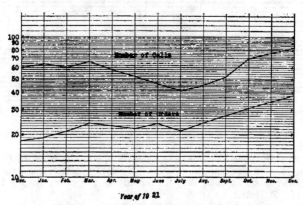

Fig. 3—Relation of Number of Orders to Number of Calls

lative fluctuations. On account of the limits of the scale, product C
cannot be read. It is therefore plotted to a larger scale as shown by
curve C. Although the relative movements up and down are apparent,
and also the general trends, yet no visualization of the changes in rate
can be made. The relative changes in rate, however, do become apparent when plotted on ratio rulings as in Fig. 5. Here it is obvious
that the downward trend of B is more marked than of A and that of
C is more pronounced than either A or B. If details are considered,

the September sales of product A in Fig. 4 on the arithmetic scale seem to have advanced the most rapidly. On the ratio scale, Fig. 5, the steepest inclination and the greatest vertical interval for that month is in product C. Contrary to the impression received from Fig. 4, product C is the one which advanced most rapidly in September.

A Graphic Chart of Interest to the sales manager is shown in Fig. 6. While this particular chart applies to the auto tire industry, its principle may be applied to industries of every description. This chart was published in *Administration*, May, 1921, in an article by Walter B. Cokell. His description of it follows:

In selling organizations of any size a large amount of data is necessary to direct intelligently the work of salesmen. The large packing houses in the Middle West, for instance, carefully compare their detailed cost figures with their sales statistics to determine which of their lines are most profitable and which among their many hundred products their salesmen can most advantageously push. It has frequently happened that the products on which handsome profits were thought

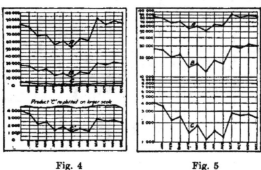

Fig. 4   Fig. 5

to have been made, when submitted to careful analysis, proved to be the very ones upon which losses were incurred.

Records of sales in various territories as made by various salesmen or of the same salesmen in the same or different territories over a period of time are often an index of efficiency. In nearly every commercial business certain statistics are also used as a measure of present efficiency or as a means of determining future possibilities. For instance, the sales of all or certain lines of a concern's product may be compared with the total population, or perhaps with certain classes of the population. Statistics showing the number of farms, by sizes, crops of grain, etc., would be useful information for agricultural machinery companies or fertilizer companies in studying their markets.

An analysis of the census of manufactories will yield a wealth of information relative to the distribution of various industries about the country and the production of most lines of manufactured goods. While some of the reports published by the federal government are suitable only for broad surveys, there are many special reports issued by governmental or private agencies which bear on individual industries. These give the results of exhaustive inquiries and indicate the location and size of manufacturing plants, the sources of supply of raw material and labor, and the extent of the market. Such information is of prime importance to a sales organization endeavoring to increase sales. Even simple statements of sales in the various districts of the field will, if rendered promptly and frequently, be a great aid to an executive in feeling the pulse of trade.

In many organizations quotas are established at the beginning of a season for the different branch offices or salesmen. The home office informs each salesman of the minimum sales of each product he is expected to make. These quotas or "shooting marks" are established by distributing over the various territories, usually on a basis of past performance, the total volume of business the concern hopes to secure. Re-

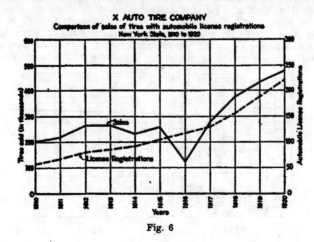

Fig. 6

ports are then issued showing how the current sales of all products compare with the quota expected to date.

Manufacturers of tires and auto accessories often compare the sales of their product in certain territories with the automobile license registration, in this way determining the size of their market and the possibilities of increasing sales therein.

Fig. 6 shows a comparison of auto tire sales with license registration. By the study of such a chart the manufacturer learns to what degree his sales have kept pace with the growth in the use of automobiles. The slump in his sales in 1914, for instance, may have been due to the inability to secure raw material on account of the breaking out of the war. This would have caused his sales in all territories to fall off. The large decrease in sales in New York State in 1916 may have been due to large war orders from abroad tying up a major portion of his output. In the study of such a chart all the factors influencing the conditions must be known if correct deductions are to be drawn from the statistics and if the figures are to be accurately interpreted.

Elaborate questionnaires are often prepared by special investigators or by the regular sales forces to gather information relative to a particular market i. e., as to the needs and tastes of the public and as to the methods of securing more customers.

In selling campaigns it is advantageous to keep statistics of the prog-

ARTICLE	PRODUCTION	SALES QUANTITY			SALES VALUE			SHIPMENTS QUANTITY			SHIPMENTS VALUE		
		TO-DAY	THIS MONTH	THIS YEAR	TO DAY	THIS MONTH	THIS YEAR	TO-DAY	THIS MONTH	THIS YEAR	TO-DAY	THIS MONTH	THIS YEAR
Standard No. 0													
No. 1													
No. 2													
No. 3													
Hot Water No. 1													
No. 2													
Units													
TOTAL INCUBATORS													
Colony Brooder													
Stove Brooder, Sr.													
" " Jr. 1													
" " Jr. 2													
TOTAL BROODERS													
K D Hoovers													
Box "													
Jr. Portable Hoovers													
Home "													
Units "													
TOTAL HOOVERS													
M 112 Cabinet													
212 "													
312 "													
412 "													
M 120 "													
220 "													
320 "													
420 "													
T 12 "													
22 "													
32 "													
42 "													
T 18 "													
28 "													
38 "													
48 "													
Special													
12-qt. Container													
20-qt. Container													
TOTAL CABINETS													
4-gal. Can													
8-gal. "													
12-gal. "													
20-gal. "													
TOTAL CANS													
12 Tub													
20 "													
TOTAL TUBS													
Parts													
Truck Bodies													
Arro Kars													
Cedar Chests													
GRAND TOTAL													

Fig. 7—Form For Report Giving Details of Sales and Shipments

ress made, to stimulate business. For instance, a corporation may pay a monthly bonus to salesmen for any excess in sales over the normal increase ordinarily to be expected from the growth of the company. The main problem is then to determine what would be the normal sales for the period if no special drive were being made. The average sales would probably be computed for several years past and on the basis of these figures a total estimated amount of normal sales for the coming year would be computed, figuring the same per cent. of increase over the previous year as had been made annually in the past. As the volume of sales in most businesses fluctuates from month to month, due to seasonal con-

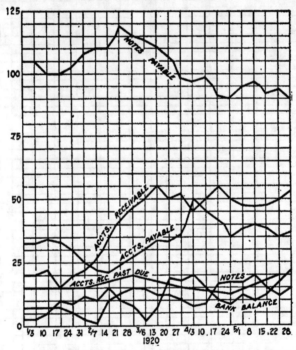

Fig. 8—Weekly Chart Compiled from Reports

ditions, it would be necessary to divide the total estimated normal sales for the coming year on the basis of the proportions which had come in each month in the past, using the results of several years as a basis. The estimated normal sales per month would have to be split up over the territory of the company, probably on the same basis as outlined above, but modified where local conditions in certain territories seemed to require more or less arbitrary adjustments. Bonuses would then be paid for any increase of actual sales above these normal figures.

H. J. Fletcher, in an article in *Industrial Management*, January 1,

1921, has given charts of production, sales and shipments and explains them as follows:

The chief executive of a modern plant at the present day is a man who has a great amount of detail constantly passing before him, and the finer that this detail can be boiled down so that it may be readily absorbed, the more promptly and accurately will he be able to make his decisions.

These details cover such data as sales, production, shipments and finance; and where a great volume of business is concerned a report should be made daily. Where the volume of these data is steady the report may be made weekly.

The form, Fig. 7, is one which is being used with success by the general manager of a large wood working plant. This report is made out daily from information supplied by the sales, accounting and production divisions. Weekly copies are compiled and sent to such members of the board of directors as are on the Finance Committee, Production Committee, etc. This report is itemized as to the various types and sizes of products. Sub totals are shown for each class of product and a grand total for the entire sheet.

On the reverse side of the sheet are spaces for financial value, such as deposits, withdrawals and balances for each bank with which an account is carried, and also the totals. Space is provided for the value of the current assets and liabilities, such as accounts and notes receivable and accounts and notes payable.

From this report a weekly graphic chart, Fig. 8, is made of the financial data, showing the accounts and notes receivable, the accounts and notes payable, the total cash in bank, and a monthly line showing the portion of accounts receivable which is past due.

Another graphic report, Fig. 9, is made of the production, sales and shipment of each class of product.

By watching the graph of the financial data the treasurer is enabled to keep a close touch on his accounting and collection department while the general manager can obtain valuable information as to the activities of his sales and factories organization.

Making Sales Statistics Talk. This was the title of an article published in the November, 1921, issue of *Industry Illustrated*. The following is taken from the article: All of us old timers remember the days of the "star salesman." As far as we could observe, the star worked about one week out of every month. His powers of Personality and Persuasion, however, were so great that he topped the fellows who kept plugging day after day.

Today very few sales managers would knowingly hire a *star* salesman. Personality and Persuasion are in demand, but only when accompanied by Persistence. The psychology of the star is such that he is more of a comet than a fixed luminary and in addition his idiosyncrasies upset the morale of the sales force.

Personality and Persuasion cannot be developed by graphical methods, but Persistence lends itself admirably to this sort of treatment, and progressive sales managers are rapidly taking advantage of it.

The principle underlying this kind of sales stimulation is the pictorial appeal. The salesman is made to see vividly exactly where he stands, either in relation to his own previous performance or in relation to the performance of other salesmen. There are so many varieties and combinations that are possible that an ingenious sales manager finds the supply almost inexhaustible.

Comparisons with a salesman's previous average or with his individual bonus peg or budget quota, are generally fair. Comparisons of

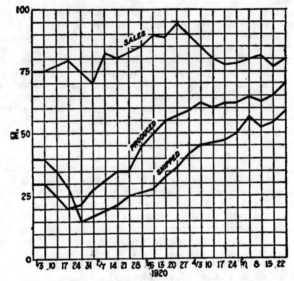

Fig. 9—Graphic Record of Production, Sales and Shipment

performance with other salesmen should only be made when conditions are equalized. It is universally realized that one territory may yield a given amount of sales while the same selling ability and energy applied in another would yield but half as much. Modern point systems and quota systems of rating and apportioning territory equalize these conditions, and where these are in effect comparisons are quite fair.

Graphic charts are unusually effective in stimulating Persistence in sales competitions. They provide this stimulant by bringing home vividly to each salesman the spirit of competition which exists in those worth while; combined with a means of recording results that helps to turn the work into play. There are so many chances for variety that the story never grows old for the sales manager can draw from the field of baseball, foot ball, sixday bicycle races, golf, track events, and even poker if need be!

There is sound business psychology back of this plan of stimulating Persistence. The old school salesmen will ridicule the idea. It is interesting to note that, however, where such plans have been introduced it is not many weeks before the old timers as well as the

younger generation are observed earnestly studying the charts and
striving to keep their records above par.

And, after all, the more that the spirit of play, honest competition

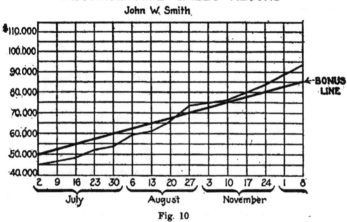

Fig. 10

Fig. 11

and interest in the score are brought into business and industry, the
more interesting will industry and business become to those who like
to "play the game."

A chart like Fig. 10, corrected up to the current date, is sent weekly
to each salesman. It shows him at a glance whether he is earning a
bonus or not, and if he is, whether the bonus is increasing or diminish-

ing. This chart is of the simple "comparative" type, the comparison being in this case against the bonus "peg." It is made to fit the sales book.

Fig. 11 shows weekly performance, the comparison being made between all salesmen, or all in a certain class or territory. The rectangle representing the salesman to whom the chart is sent is put in black, so he can quickly visualize his performance as against his fellow

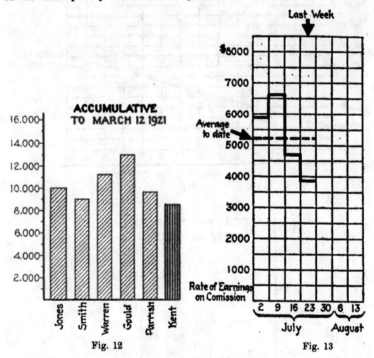

Fig. 12   Fig. 13

salesmen. Fig. 12, on the same principle, shows accumulative sales since the first of the year.

Fig. 13 is another simple "comparison" chart for salesmen. In this case the comparison is last week's performance against the average to date for the year, half year or quarter. The arrows are put on with rubber stamp, and a new chart is mailed each week. This chart tells the commission salesman at a glance just what annual salary rate he is earning.

Sometimes the graphic record of a competition provides a first-rate basis for a sales letter. Fig. 14 is an example which has the airplane as its motive and which is very effective in its appeal. Any single motive of this sort kept in operation over too long a period, will grow stale, and there is no excuse for this happening in view

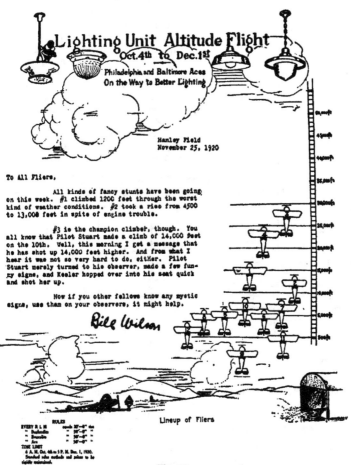

Fig. 14

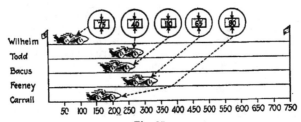

Fig. 15

of the many varieties of motives available for graphic use.

Fig. 15 is a graphical competition chart with a double purpose. The distance of each racing car from the starting line indicates the cumulative sales points or dollars while the speedometers above indicate the "rate" of travel for the last period. Thus Feeney, although in the lead, sees that he will have to put on gas to keep Bacus from catching him with his 110-mile clip. Charts of this sort elaborated in scale and planned for the home office, stimulate interest among all employees whether salesmen or not, and add a secondary urge to the salesmen to excel.

Where the effort is to develop salesmen rather than to rank accomplishment by gross results, a graphic chart like Fig. 16 shows at a glance which man is making the greatest per cent. of improvement.

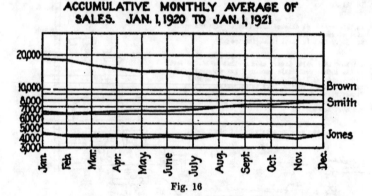

Fig. 16

Thus Brown, who has the best total sales, is nevertheless going down hill fast, and Jones is plodding along on the level, but Smith is traveling up-hill and appears to be a comer.

"Curves That Plot the Performance and Efficiency of the Salesman," following, was published in the *Electrical World*, August 12, 1916:

The Elmira Water, Light & Railroad Company of Elmira, N. Y., employs a graphical method for following the work of its salesmen. As explained by F. H. Hill, general manager of the company, the salesmen's efforts are reduced to a curve, Fig. 17.

This particular salesman, it will be noted, is an extremely active man, making a very large number of calls per day during the month for which the curve was compiled. The percentage of his sales and his calls, however, is also quite high, and the cost per sale is extremely low. By the word "sale" is meant the securing of a meter contract. The man whose work is indicated on the curve sheet is exclusively on residence work.

At the end of each month each salesman's report is made up on this form and each salesman is given a copy, not only of his own but of every other man's record. "We find," comments Mr. Hill, "that each salesman spends more time in studying the other man's record than he does his own, endeavoring to find out how his competitor does it."

Knowing Just What Your Salesmen Are Doing. Mr. Stanley C. Tarrant, in the course of an article published in *System, the Magazine of Business,* December, 1915, entitled "Graphs for the Sales Manager," tells how a sales manager uses charts to keep a record of his salesmen. See Fig. 18.

The chart is drawn on millimeter paper. A six-foot length of this paper takes care of the records of 27 salesmen. It shows graphically 7 facts about every salesman. In the upper portion, the following facts appear:

1. Number of orders obtained: black line;
2. Value of orders obtained: black dotted line;
3. Number of calls made: green line;
4. Average value per order: green dotted line.

On the lower portion of the chart, these facts are shown:

1. Average cost per order (salary, commission and expenses): black line;
2. Average cost per call: black dotted line;
3. Average cost per $100 of orders: green line.

These 7 points are plotted and drawn on the one chart to show the records of 27 salesmen in less than 4 hours a month.

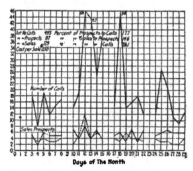

Fig. 17—The Efficiency of the Salesman

To secure the information which goes on the chart, each salesman is provided with a supply of post cards. He fills out one of these every day and mails it to the office. On it, in the convenient spaces it provides, he shows the number of calls he made during the day, the number of orders he obtained and the value of these orders.

When the cards are received at the office, the information each carries is transferred to the salesman's individual summary and at the end of each month the figures on this form are totaled and the averages figured.

Instead of the large sheet showing the record of the 27 salesmen another very satisfactory method would be to have a separate chart, 8½ x 11 inches in size, for each salesman. These could be filed in alphabetical order in a loose-leaf book.

To Interest Salesmen and Increase Sales. A form of chart intended to excite interest and speed up sales is shown in Fig. 19. This and the following article, "Taking Salesmen's Selling Temperature," appeared in *Motor World*, April 25, 1917. The Grasser Motor Co., Hupmobile dealer, Detroit, was speeding up spring sales by holding a salesman's contest during the month of April.

First, the branch manager estimated how many cars ought to be sold in the month. He knew that all his salesmen were not equal, and hence could not sell proportional parts of this estate. So to put the contest on a fair basis, and prevent discouragement, he set a definite number of cars the men must sell to qualify for a prize.

To win first prize, $100, the salesmen must sell at least 12 cars; to win second prize, $75, 10 cars, and third prize, $50, 7 cars.

The record of the salesmen is kept on a chart, the amounts sold being shown by the height of the column in the thermometer representing the individual salesman's sales.

In order to prevent the men holding back on their sales and to in-

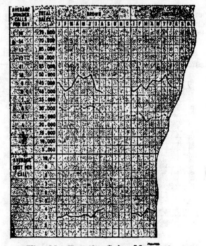

Fig. 18—For the Sales Manager

duce them to make a good record through the contest, he did not announce the contest until April 1. Hence the men all started equal.

The contest was preceded by an advertising campaign that tended to plough up the prospect field and stimulate the sales, but the main effort was concentrated in creating a sporting spirit among the salesmen themselves. The sales contest did this.

Keeping the Cost of Selling Down. Edward Corrigan in an article "What it Should Cost to Sell," published in *System, the Magazine of Business*, October, 1917, tells of the method his concern has adopted to standardize and reduce selling costs. The following is taken from this article:

In the first place sales work is based upon an intangible commodity: human nature, in both salesman and customer. It is impossible to know exactly to the penny whether a salesman is producing at the minimum cost in a given territory. Again, some buyers are quick to decide while others will take a day to transact business that could be finished in an hour. Some customers are small buyers, and it is necessary to see them frequently; others buy heavily and need to be visited only two or three times a year.

We found, therefore, that we could devise no hard and fast system of cost, as we could in the factory. All that we could do was to work out a fairly accurate system.

Our first step was to fix for each territory what we considered an equitable standard of cost. In one territory the standard might be double that in another. Where selling costs by territories were already kept, they formed the basis of the standard. Where the costs as a

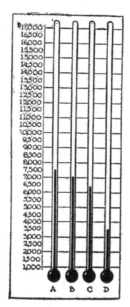

Fig. 19—Salesman's Selling "Temperature"

whole only were available, it was necessary to work out an approximate cost for each territory. As the cost system developed, we adjusted the standard for each territory.

Unit of product, we found, afforded us the best basis in fixing this standard.

In addition to the standard selling cost, we established a quota of sales for each territory.

In figuring the actual cost, we give the salesman credit for all business from his territory, whether he actually sends in the order or we receive it by mail.

In fixing both standards of sales cost and quotas, we felt justified in figuring the cost a little low and the quotas a little high. Of course, we had to be careful not to put the cost too low, or the quota too high, and thus discourage the salesman.

We try always to keep the idea of reducing the unit cost of sales before each salesman in the most convincing form. Any increase in the quota, of course, is a step tending toward the reduction of sales cost. We plot curves to visualize the facts. In Fig. 20 the heavy line is the

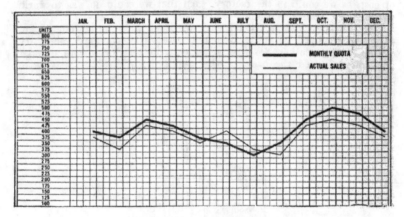

Fig. 20—Keeping Up to Quota

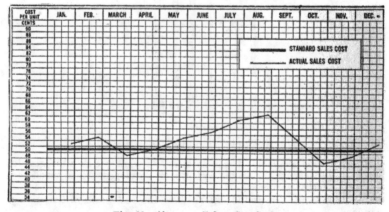

Fig. 21—Above or Below Standard

quota for a territory covering a period of twelve months, and the light line shows the actual sales in the same period.

With these figures so graphically before us, it is simple to make up new quotas each year. We usually plan for a certain percentage of increase in sales over the same months of the previous year. You will notice in the curve shown that allowance is made for the usual dull months of summer. As a rule, we find that a standard of sales cost or a sales quota should be based on the same month of the preceding year rather than the previous month of the same year.

Fig. 21 shows the standard sales cost for a certain territory. The heavy line represents the standard cost fixed for the territory, and the light line is the actual cost.

In both of these charts the lines representing the quota and the standard cost are drawn in when the charts are made up at the beginning of the year. At the end of each month the curves showing the actual sales, or the actual cost to date, are drawn in and sent to the salesman.

What and When People Buy. The following is from an article by Theodore G. Morgan, entitled, "I Discover Why My Store's Advertising Pays," published in *System, the Magazine of Business,* January, 1916. Mr. Morgan secured the assistance of the head of the psychological department of a university who gave him the theories and between them they worked out some practical and profitable results:

Our idea was that if we could define the habits of the shopping public, we could time our appeals to appear at the moment when they would be most effective. If we could do this we might greatly reduce such waste as had occurred in our advertisements.

So we made a careful study of the various departments, by personal observation, by reports from managers and salespeople, and through records of actual sales. What we wanted to do was find on what days and at what seasons the greatest number were in the habit of purchasing particular lines.

We uncovered many interesting facts. For example, we learned that materials sold by the yard, such as silks, dress goods, trimmings and wash fabrics, were purchased on Mondays and Tuesdays in larger quantities than on any other days of the week. The most opportune time to present announcements of these materials, therefore, was naturally on Monday and Tuesday. By doing this, our test showed that we greatly increased the pulling power and reduced waste.

In the same manner we discovered when the greatest number of people purchased dress patterns, trunks and bags, and wedding gifts; and we classified and rescheduled our appeals in these lines, as well as in other lines, to reach the public at times when habits made the greatest number of people naturally think of these particular articles.

The time to advertise wedding gifts for June weddings, for instance, is not in the middle of June, when the weddings are actually taking place, but during the month of May when people are receiving wedding invitations and deciding what they ought to give.

Likewise, the time to advertise fine trunks is not when the vacation season is in full swing, but from three weeks to a month previous, when preparations for travel are being made.

Careful study showed us, again, that the public expects bargains on Friday, and may come to stores on that day with no particular aim except to hunt bargains. We recognized the importance of this habit, and made it a point to place our incomplete assortments on sale on that particular day each week.

It was easy for us to note, further, that on Saturdays there were

more children in the store with their parents than on any other day
in the week. The answer was simple—Saturday was the only day in
the week that the children were free from school. By assembling all
our children's advertising in the Friday papers we reached the homes
when mothers were wondering where they should go to purchase things
for their children the following day.

We noticed, too, that furniture is most keenly sought on Saturday
afternoons and on Mondays. Saturday afternoons the husband can accompany his wife. The reason for Monday buying, we determined, was
the Sunday holiday, during which the husband and wife can discuss
the desirability of so important a purchase as furniture. It is perhaps
the most important item purchased for the home, and is not usually
bought without deliberation. Furniture appeals reaching the public,
then, for Saturday and Monday shopping, were directly in accord with
a clearly defined habit.

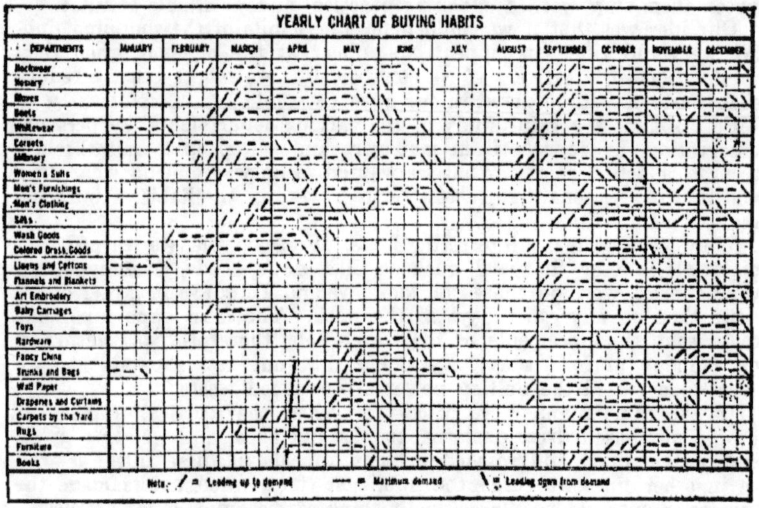

Fig. 22

Still another discovery: there is great activity in ready-to-wear ap-
parel and dress accessories during the latter part of each week. Many
buy then in preparation for a week-end outing, or in order to meet
the requirements of Sunday. And it requires so little time to select
these things, that purchasing them can safely be put off until the last
moment.

In carrying out our plans, we first of all made a large "chart of demand." This was a visual chart, based upon the experience of former
years. It enabled us, at a glance, to determine the volume of future
demand, and to plan and prepare our advertising appropriation to
coincide to the best advantage.

The secret of successful buying and selling lies in judging this trend
and strength of future demand. With our chart, the merchandise de-

partment knows in just how many weeks demand will be at its maximum for a given line, and buying is permitted or curtailed accordingly.

For instance, actual merchandise figures show that a department is overbought. The manager of that department is anxious to go into the market for several new lines, and argues that a four weeks' season is just beginning, which will develop a strong demand for the new line. Often a buyer, in his desire to buy, is overoptimistic in this way. But the chart is a definite restraining guide, Fig. 22.

It will be seen how this shows the length of maximum demand, and how permission to buy can be given or refused largely by finding how long a period of maximum demand still remains.

The Use of Maps in Graphic Sales Analysis. Outline maps, Fig. 23, are very valuable in aiding the sales manager to visualize the progress of events in his department. Upon such maps may be shown

Fig. 23—Outline Map of United States

the sales by territories. Either different colors may be used to designate the sales, such as a red dot for 0–$1,000, a blue dot for $1,000–$2,000, etc., or else the actual figures of the sales may be written directly upon the map. Where various products are sold, the sales of each different kind may be shown according to locality and will indicate how the locality affects the sales of each kind.

In some businesses population may be taken as a basis of sales. This would be the case where the product had a universal appeal and was not restricted to any specific class. Then the map may be used to show the ratio of sales to population for various districts and affords an excellent basis for concentrated advertising. Where the demand for a product is limited to a certain class it may be indicated upon the map what the estimated number of prospects is in various localities.

The outline map is also useful for charting general information about each state, the desired data being marked right on the map and following some predetermined sequence, such, for example, as: (1) total population; (2) percentage of population engaged in agricultural pursuits; (3) in manufacturing; (4) number of cities with population below 10,000; (5) from 10,000 to 25,000; etc.; (6) miles of railroad; etc., etc., depending upon the facts that are valuable. Of course, when entering the data on the map the reference figures only need be used, such as: (1) 2,580,000; (2) 43 per cent.; etc.

The maps may be used to allocate to various territories the number of salesmen estimated to be necessary to handle the probable business. The routes of salesmen may also be shown, with the dates when they are scheduled to arrive at and depart from various places along the routes.

CHAPTER XXXIV

Graphic Charts Applied to Scheduling and Production

There is hardly a successful business being conducted today which does not have some method of production control. Where there are so many different processes entering into the manufacture of a finished product, it becomes absolutely necessary for the manager to know definitely just how long it requires to carry out each process if he is to coördinate them efficiently.

As a means of accomplishing this result there have been evolved many different systems. In most of these the use of graphic methods plays an important part. The reason for this is because charts will give a comprehensive picture of what work has been planned, what has been done and at what rate it has been done. They may be made to show the reasons why actual performance has not kept pace with scheduled performance, and they furnish a means of directly fixing the responsibility for such failures. They make it possible to schedule work so that all of its various processes will coordinate and they furnish a basis for accurately estimating the time required to do new work which involves these processes, all or in part. The charts may be simply made, so that if it seems desirable to show them to the workmen they can easily understand them.

The Gantt Chart. The Gantt chart was originated and developed by H. L. Gantt for graphically showing the relation of the actual performance to the expected performance of a man, machine, shop or plant. It is primarily designed to keep an executive in close touch with conditions by giving him a picture of what is actually happening. In this it is no different from charts in general, but it presents a number of advantages for scheduling, production and progress.

A comprehensive series of articles on the Gantt chart has been written by Wallace Clark in the August, September, October and November, 1921, issues of *Management Engineering* from which the following notes have been taken.

In the Gantt chart a division of space represents both an amount of time and an amount of work to be done in that time. Lines drawn horizontally through that space show the relation of the amount of work actually done in that time to the amount scheduled. Equal divisions of space on a single horizontal line present at the same time:

1. Equal divisions of time.
2. Varying amounts of work scheduled.
3. Varying amounts of work done.

Thus it shows the relation of time spent to work done. Furthermore, since knowledge of what has happened and when it happened causes action, the past projects itself into the future and records charted in this way become dynamic. A single example may make this method clear.

A week's work is planned as follows:

Monday 100, Tuesday 125, Wednesday 150, Thursday 150, Friday 150. A sheet is ruled with equal spaces representing days (Fig. 1) and the amount of work planned is shown by figures on the left side of the day's space. So far the chart shows the schedule and its relation to time.

The work actually done through the week was:

Monday 75, Tuesday 100, Wednesday 150, Thursday 180, Friday 75.

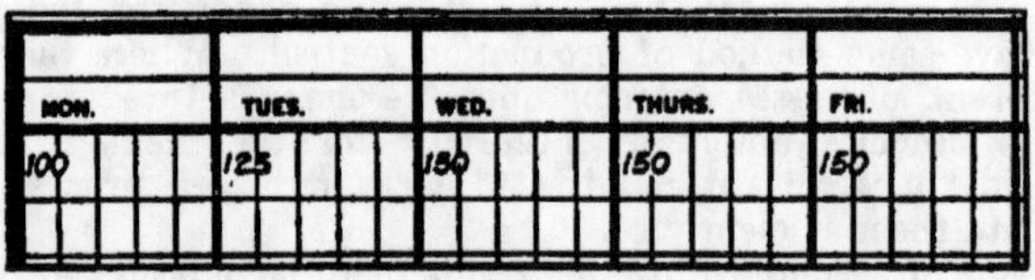

Fig. 1—Gantt Chart Showing the Daily Schedule

This is charted as shown in Fig. 2.

Lines are drawn through the daily spaces to show a comparison between the schedule and the actual accomplishment. On Monday the space represents 100; only 75 were done, so a light line is drawn through 75 per cent. of the space. On Tuesday 125 were planned; 100 were

Fig.2 —Gantt Chart Showing the Work Actually Accomplished

done; a line is therefore drawn through 80 per cent. of the space. On Wednesday 150 were to be done and 150 were done, so the line is drawn through the entire space. On Thursday 150 were scheduled and 180 were done, i. e., 120 per cent. of the schedule; a line is therefore drawn all the way across the space to represent 100 per cent. and an ad-

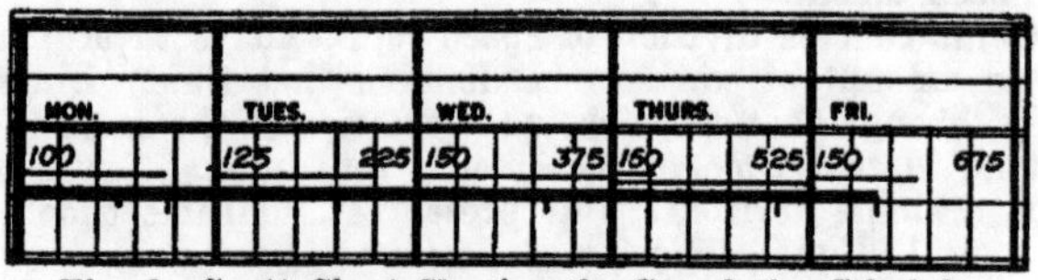

Fig. 3—Gantt Chart Showing the Cumulative Schedule and the Cumulative Work Done.

ditional line through 20 per cent. of the space. On Friday 150 were planned; but only 75 were done; a line is accordingly drawn through 50 per cent. of the space. The chart now gives a comparison day by day of the amount of work done and the amount scheduled and the relation of both schedule and accomplishment to time.

It is, however, desirable to know how the whole week's work compares with the schedule and so the figures representing the *cumulative schedule* are entered on the right side of the daily space (Fig. 3). At the end of the day on Friday, for instance, the total amount to be done up to that time was 675. A heavy line is therefore drawn to show a comparison between the *cumulative work done* and the *cumulative schedule*. On Monday the heavy line is the same length as the light line. Of the 100 done on Tuesday, 25 have to go to make up the shortage for Monday. The remaining 75 are applied on Tuesday's schedule and the heavy line is drawn through 60 per cent. of the Tuesday space. Of the 150 done on Wednesday 50 are needed to meet the schedule to Tuesday night and the remaining 100 are applied on Wednesday's schedule of 150, the line being drawn through 66 per cent. of the space. Of the 180 done on Thursday 50 are used to meet the schedule to Wednesday night and the line representing the remaining 130 is drawn through 87 per cent. of the day's space. Of the 75 done on Friday 20 go to meet the schedule to Thursday night, leaving 55 to be applied to Friday. The cumulative line, therefore, shows us that on Friday night the work is two-thirds of a day behind the schedule. This chart (Fig. 3) shows the relation of the schedule to time; the work done each day in relation both to time and the schedule, and finally the cumulative work done and its relation to time and the schedule.

Entering the Schedule. At the top of the sheet (see Fig. 4) enter a description of the information to be charted on the sheet, placing at the extreme left the one or two words which distinguish this sheet from others in the same binder. At the heads of the columns representing units of time enter the dates. In the columns on the left side of the sheet write a description of the work to be charted on the various lines.

The date or hour when work is to be begun is indicated by a right angle opening to the right, thus: ⌐

The date on which this is to be completed is indicated by an angle opening to the left, thus: ⌐

The amount of work scheduled for any period of time is indicated by a figure placed at the left side of a space, thus:

| 10 |

The amount of work to be done up to any specified time is indicated by a figure placed in the right side of a space, thus:

| 40 |

If these entries are made by hand, use India ink so that good blue-prints can be made. If they are typewritten use a heavily-inked black ribbon, and place a sheet of carbon face up against the back of the paper. The resulting blue-prints will show clear white typing.

Entering Work Done. Light lines represent work done during any given period of time, thus:

|——————————— |

The length of line bears the same relation to the width of the space as the amount of work done bears to the amount scheduled. Heavy lines represent the cumulative amount of work done and show its relation to the amount scheduled to be done up to a given date. Thus:

▌═══════════ |

The Uses of the Gantt Chart. The Gantt Chart shows facts in their relation to time, emphasizing their movement through time, and therefore it compels a man to take action based on the facts shown, just as if he were responding to a force of motion.

The use of a Gantt Chart makes it necessary to have a plan; it compares what is done with what was planned; it shows the reasons why performance falls short of the plan; it fixes the responsibility for the success or failure of a plan; it is remarkably compact; it is easy to draw and easy to read; it visualizes the passing of time, and therefore helps to reduce idleness and waste of time; it measures the momentum of industry.

The Layout Chart helps to plan work so as to make the best possible use of the available men and machines and also to so arrange orders as to secure whatever deliveries may be desired. The Load Chart keeps executives informed as to the amount of work ahead of their plant and enables them to co-ordinate workmen, equipment, processes, orders, and prices. The Progress Chart helps to get work done by showing a comparison of what is done with what should have been done and enables the executive to foresee future happenings with considerable accuracy. It shows the effect of past decisions and points out the action which should be taken in future.

The Man Record and the Machine Record Charts show whether the management is good or bad.

Are the machines being run all day?

Are the men doing a full day's work?

If not, what are the reasons?

The answers to these questions are the ultimate facts in regard to the management of any manufacturing plant. If the men and machines are doing a full day's work, it is obvious that all the other details in the management of the plant are being taken care of on time. Shop orders, production cards, layout charts and reports of all kinds are merely a part of the mechanism which leads up to or follows the Man and Machine Record Charts. They measure the service rendered by the workman, the foreman, and the management.

Each day when a man fails to do a fair day's work the reason is shown on a Man Record Chart. If that failure is due to absence, slowness or avoidable mistakes, it is the fault of the workman, but if his failure is due to lack of instruction, to tool troubles or to a machine in need of repairs, the fault is with the management. This chart, therefore, measures the service rendered by the individual workman and also the use the management makes of his service. The Machine Record Chart measures the ability of the management to make satisfactory use of the equipment at its disposal. When a machine is not running the reason for its idleness is shown, and from the reason the responsibility can easily be traced.

Gantt Machine Record Charts. The purpose of the Machine Record Chart is to show whether or not machines or equipment are being used and, if not, the reasons for idleness. In a manufacturing plant the foreman uses a sheet ruled to represent the working hours of his shop or department. If he works an 8-hr. day he has each wide column which represents a day ruled off into four narrower columns each representing two hours. If he works a 9-hr. day, he rules the day off into four wide spaces of two hours each and one narrower space for one hour.[1]

On the left side of this sheet the foreman or his assistant lists all the machines, benches, or work spaces in his department, arranging

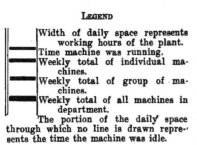

Fig. 4—A Gantt Machine Record Chart

LEGEND

Width of daily space represents working hours of the plant.
Time machine was running.
Weekly total of individual machines.
Weekly total of group of machines.
Weekly total of all machines in department.

The portion of the daily space through which no line is drawn represents the time the machine was idle.

REASONS FOR IDLENESS

E—Waiting for set-up   P—Lack of power
H—Lack of help   R—Repairs
M—Lack of material   T—Lack of tools
O—Lack of orders   V—Holiday.

When there is more than one reason for idleness, the reason entered on the chart is determined by asking questions in the following order:

M—Is the machine ready to run?
O—Is there an order for the machine?
M—Is there material ready to be worked on?
T—Are there tools?
P—Is there power to run the machine?
H—Is there an operator for the machine?

them in groups according to responsibility, if there are any subforemen. If not, they are arranged by kinds of machines. At the top of each group he leaves a space for the total of that group and at the top of the sheet a line for the total of the department. (Fig. 4.)

[1] The Codex Book Co., Inc., New York, carry various forms of the Gantt Chart in stock.

Opposite each machine number the foreman indicates whether or not the machine has been running by drawing a light line across the space to indicate how many hours the machine ran. The ratio of the line to the space is the same as the ratio of the hours the machine ran to the working hours of the plant. A blank space indicates that the machine did not run, and in that space a letter or symbol is placed to indicate the reason why. The letter indicating the reason is placed at the beginning of the space representing the idleness so that it would be bisected by the light line if there were any.

Under the light line a heavy line is drawn to indicate the cumulative running time of the machine for the whole week. The length of this heavy line is always equal to the sum of the light lines for the various days. The heavy line rests on the printed line and the light line is drawn about 1/10 in. above the heavy line.

The running time of the individual machine in a group is averaged and the light and heavy lines entered for the group total. In the same way the groups are averaged to get the total running time of the shop, and the lines are drawn at the top of the sheet.

In this Machine Record Chart the foreman has a graphic record of the running of his machines which enables him to visualize his problem and to grasp the facts and the tendencies much more firmly than he could from any written record or from watching the machines. Moreover, the chart emphasizes above everything else the reasons for idleness of machines, and so it indicates very clearly who is responsible for the idleness.

Since it is the foreman's aim to get work done, he studies the facts shown and translates the chart into action. He eliminates as much as possible of the idleness over which he or his subordinates have control. If machines have been "waiting for set-up," he plans the work of his set-up men more carefully, and, if necessary, trains an additional set-up man. If the machines are idle for "repairs," he does all he can to push the completion of the repairs. If the trouble is "lack of material," he asks the storekeeper for help.

A considerable part of the idleness of machines appears to be due to causes over which the foreman has no control, so he takes the matter up with his immediate superior, who may possibly be the superintendent.

In order to get a better idea of the progress made in the running of his machines, the foreman prepares a Summary of Idleness Chart (Fig. 5), on which he enters each week the one line which summarizes his whole department and he shows the hours of idleness due to the various reasons.

The foreman in whose office these charts are kept not only advances his own interests by keeping them, since they enable him to become a more important and capable man in the eyes of the management and of his workmen, but by the same means he calls to the attention of the other individuals their responsibilities.

The Machine Record Charts are of great value to the superintendent because they bring to his attention the problems on which his help is most needed. He does not have to go around the shop asking his foremen what is wrong and frequently finding out only when it is too late. The obstacles which prevent his foremen from keeping their machines running are brought to his attention regularly and in detail. In order to get a comprehensive grasp of conditions, he has the records of all his departments summarized on an Idleness Expense Chart.

Because of his greater experience and broader authority the superintendent can be of most service in advancing production by helping the foremen overcome the obstacles with which they are daily confronted and which they report to him on the Machine Record Chart.

Gantt Man Record Charts. The purpose of the Man Record Chart is to show whether or not a man does a day's work and, if not, the reason why. In keeping a Man Record Chart the foreman uses a sheet which is ruled according to the working hours of his shop and is similar to the one used for the Machine Record Chart. On the left side

PRODUCTIVE MACHINES	PER CENT OF CAPACITY USED 10 20 30 40 50 60 70 80 90	TOTAL HOURS OF IDLENESS	DETAILS OF IDLENESS HOURS DUE TO							
			LACK OF HELP	LACK OF MATERIAL	LACK OF ORDERS	LACK OF POWER	REPAIRS	LACK OF TOOLS	WAITING FOR SET-UP	HOLIDAY
Week Ending July 5th		1972	236	302	29	0	341	178	62	764
" 12		1478	252	436	260	0	333	152	48	0
" 19		1675	241	471	387	0	402	138	36	0
" 26		1478	206	523	115	88	357	147	42	0
August 2		1421	192	437	318	0	328	120	26	0
" 9		1336	180	413	318	0	303	104	18	0
" 16		1309	186	387	331	0	281	116	8	0
" 23		1205	173	307	336	0	294	83	12	0
" 30		1035	164	324	130	148	268	61	0	0
Sept. 6		1319	96	282	257	0	239	40	0	405
" 13		1151	168	253	154	0	178	29	0	369
" 20		873	101	206	340	0	191	35	0	0
" 27		882	83	263	460	0	64	12	0	0
Oct. 4		777	48	241	345	0	143	0	0	0
" 11		760	22	213	468	0	57	0	0	0
" 18		1035	35	178	416	0	38	8	0	360
" 25		815	16	192	558	0	49	0	0	0
Nov. 1		610	9	157	382	0	62	0	0	0
" 8		865	0	106	352	0	47	0	0	360
" 15		527	5	85	382	0	55	0	0	0

Fig. 5—A Summary of Idleness Chart

of the sheet he lists the men in his control arranged in groups under his sub-foremen, if he has any.

If the operator has fallen behind in the work expected of him, the reason is indicated by a letter (see Fig. 6).

The foreman watches the first line of the chart because it shows him how his department as a whole is living up to his idea of what it should do. The foreman is usually surprised to see that the failure of the operator to do the work within the estimated time is more often his fault than that of the workman. He learns how much of the time of his men is wasted because of the improper sharpening of tools, defects in materials which should have been caught by the inspectors, the unsatisfactory condition of machines and the lack of proper instructions on new work. He understands better than ever before why the costs of so many jobs exceed his estimates.

In order to get the help of the superintendent in removing delays over which he himself has no control, the foreman sends copies of his

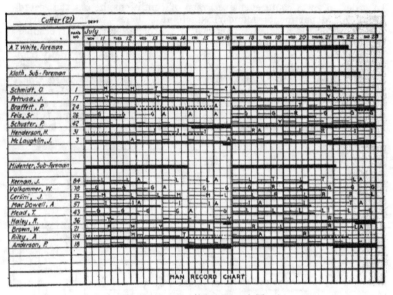

Fig. 6—A Gantt Man Record Chart

REASONS FOR FALLING BEHIND

A—Absent R—Repairs needed
G—Green operator T—Tool troubles
I—Lack of instruc- V—Holiday
tions Y—Smaller lot than
L—Slow operator estimate is based
M—Material trou- on.
bles

When there is more than one reason for failure to do the work in the estimated time, the reason entered on chart is determined by asking questions in the following order:

R—Was the machine in good condition?
T—Were the tools and fixtures in good condition?
I—Was the operator given proper instructions and sufficient information?
M—Was trouble experienced with material?
G—Was the operator too green to do the job?
L—Was the operator too slow?
Y—Was the lot smaller than the estimate is based on?

LEGEND

❙ ❙	Width of daily space represents amount of work that should have been done in a day.
————	Amount of work actually done in a day.
- - - - -	Time taken on work on which no estimate is available.
▬▬ •••	Weekly total of operator. Solid line for estimated work; broken line for time spent on work not estimated.
▬▬ •••	Weekly total for group of operators.
▬▬▬	Weekly total for department.

The portion of the daily space through which no line is drawn how much the man has fallen behind what was expected of him.

Man Record Charts to the superintendent each week. With them he sends a Man Record Summary (Fig. 7).

The Gantt Layout Chart. Idleness of men and machines is usually the greatest waste in a manufacturing plant and yet when the reasons for idleness, such as lack of help, material, orders, tools, etc., are presented to the management in such detail as to fix responsibility, it is possible to take definite steps to prevent its continuance. This is done by planning work sufficiently far in advance to advise each individual concerned what he is to do and when he is to do it. In some plants where a uniform product is manufactured, this is not a difficult matter. If, for instance, 100 machines are being made each week, every foreman or workman knows that he is to turn out enough parts to make 100 machines. The planning in such a case is very simple and can sometimes be done without any written record. There are very few plants, however, which produce only one article—usually a department has to turn out a great many different parts to be used in the

FOREMAN	PER CENT OF CAPACITY USED	TOTAL HOURS OF IDLENESS	ABSENCE	GREEN OPERATORS	LACK OF INSTRUCTIONS WORK	SLOW OPERATOR	MATERIAL TROUBLES	REPAIRS NEEDED	TOOL TROUBLES	SPOIL LOT	
Week Ending March 26th		716	151	45	146	49	37	278	2	8	
April 2		682	163	40	142	48	39	251	3	6	
" 9		654	185	40	167	44	35	182	1	0	
" 16		576	146	32	130	44	37	175	21	11	
" 23		600	134	29	123	26	40	229	8	14	
" 30		408	154	16	99	38	25	72	4	0	
May 7		473	116	28	113	47	43	120	2	4	
" 14		420	132	22	126	42	44	52	0	2	
" 21		435	107	18	81	52	44	90	29	7	
" 28		286	91	14	63	31	36	44	1	0	
June 4		355	113	12	52	36	39	76	19	6	
" 11		327	104	10	18	44	42	32	7	0	

Fig. 7—A Man Record Summary Chart

assembling of a varied product. It is also probable that these parts are worked on in other departments. It therefore becomes necessary for the foreman to plan carefully the work to be done on each machine in his department and also for the superintendent or manager to plan the work to be done in all the departments of the plant.

An angle opening to the right, as shown by the legend to the Layout Chart (Fig. 8), indicates when the job is to be started; an angle opening to the left when the job is scheduled to be completed; and a light line connecting the angles the total time scheduled for the order.

The machine on which the next operation is to be done is looked up on the chart to see when it will be ready for additional work. The order is then assigned to this machine and the angles and the light line are drawn. This procedure is followed in laying out all the operations on that order and is continued until all the orders are laid out.

In assigning work to machines, it is necessary to know what progress has been made on the work already assigned. Accordingly, as daily reports are received showing the amount of work done, a heavy line is drawn under the light line.

If the work is exactly on schedule, the end of the heavy line will be

directly under the proper date and hour. If the work is behind or
ahead of schedule, the end of the heavy line will be behind or ahead
of the date. In assigning a new order to a machine, if the work is
ahead of schedule, the new order is placed over the old one, and the
date of beginning is placed in advance of the date of completion of the
old order. The "V" indicates the date. The work is one day ahead
of schedule, and conditions in the shop indicate that it will be one day

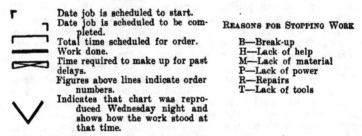

Fig. 8—A Gantt Layout Chart for a Machine Shop

LEGEND

⌐	Date job is scheduled to start.
¬	Date job is scheduled to be completed.
⊏⊐	Total time scheduled for order.
▬▬▬	Work done.
⋈	Time required to make up for past delays.
	Figures above lines indicate order numbers.
V	Indicates that chart was reproduced Wednesday night and shows how the work stood at that time.

REASONS FOR STOPPING WORK

B—Break-up
H—Lack of help
M—Lack of material
P—Lack of power
R—Repairs
T—Lack of tools

late finishing. The new order "A424," is therefore laid out to be
started Thursday morning.

If the work is *behind schedule,* there is no advantage in planning to
begin the new order until the old one is complete. Therefore, sufficient
time must be set aside to make up for past delays before the new work
can be begun. This is indicated on the chart by connecting the angles
by crossed lines. On the date shown by the "V" the work was one

day behind schedule. Before assigning order "A426," one day is allowed to make up for the delay and is indicated by crossed lines. Above the light lines are written whatever numbers and quantities may be necessary to identify the orders.

When work stops on any order a jog is placed under the line with an initial to indicate the reason. The more common ones being, as shown by Fig. 8 repairs, lack of help, material, power, or tools.

The Gantt Progress Chart. Usually it is unnecessary for the higher executive to follow on progress charts all the details of the work being done under his direction, but he does wish to know the progress of the work as a whole, and this may be done by following key operations, typical items, or totals. If the progress made on one of these subdivisions of the work is satisfactory, he will pay little attention to it; on the other hand if another part of the work is behind schedule, he will call for the detail charts in the hands of one of his subordinates. From these records he can see what particular items are being delayed and the reasons.

The Progress Chart of Crank Handles (Fig. 9) was drawn in a plant where all orders were charted. The angle opening to the right indicates the date on which the material was to be issued from stores; the figures indicate the dates on which the various operations were to be begun, that is, on the first line of the chart "1" indicates that the first operation was to be begun on January 19, operation No. 2 on the 21, etc.; the angle opening to the left indicates the date on which the parts were to be shipped; the heavy line shows what operations have been done; and the letters under the lines indicate the reasons for delay.

The "V" indicates that this chart was reproduced on March 3. If the work had proceeded exactly according to schedule, the heavy lines would all end under that date except for those orders which were due to be completed at earlier dates. However, the work had not made the expected progress; the third order on the sheet was a week behind schedule, the fifth about two weeks behind, the sixth two weeks ahead, and the seventh ten days behind time.

From this chart the manager could see at a glance which orders were behind schedule. On the fifth order on the sheet, for instance, he could see that the eleventh operation had been begun but not finished, and the "R" showed that the delay was caused by repairs. Reference to the shop order told the manager what that operation was and the department in which it was being done. Over the telephone he found out in detail from the foreman the repairs needed and the probable date when the operation would be finished. The chart showed how much more time would be needed for the remaining operations, so that the manager could take whatever action he deemed wise to rush the work and he could advise his customer as to probable delivery.

The Gantt Load Chart. The purpose of the Gantt Load Chart is to keep the executives of any producing plant advised as to the load of work ahead of their plant. This information is of particular value to managers, superintendents, foremen, employment departments and sales departments, for it gives them an accurate picture of the work which is to be done, and it is necessary to have a clear understanding of what is to be done before effective steps can be taken to do it.

The Load Chart is similar to the Layout Chart in that it shows how much work is to be done, but it is more compact than the Layout Chart and does not show details. Layout Charts show each operation on each order and the individual machines which are to do the work, but a Load Chart merely shows classes of machines and the hours of work assigned to them by weeks or months. The drawing of the Load Chart is similar to that of the Progress Chart so far as light and heavy lines are concerned; but the similarity ends there, for the Progress Chart shows work done and lines are added as more work is done; but the Load Chart shows only work which is *to be done* and represents the status of plans at a specified date. It is not a record added to day by day, but an analysis of a situation at a given moment.

At the left of the sheet are listed the classes or groups of operators, machines, work benches or floors, and in the next column the numbers in each group. In the columns representing months or weeks, the

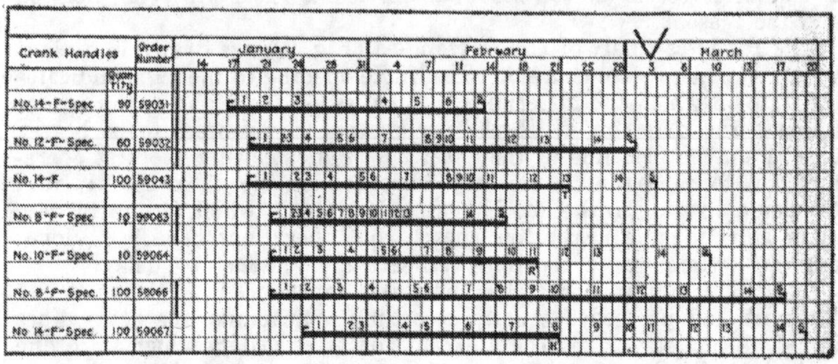

Fig. 9—A Gantt Progress Chart Used in a Plant which Manufactures on Order

figures indicate the number of operating hours for a group of men or machines; the light lines show the hours of work which have been assigned to that group during each week or month; and the cumulative lines represent the total hours of work ahead of each group. The information for this chart is secured from Layout Charts which show what orders are ahead of each machine, and from this it is easy to foot up the hours of work planned for the various classes of machines for each week or month.

When the amount of work ahead of a plant is placed before an executive on a Load Chart, it is possible for him to so grasp a situation that he can adjust equipment, operators, and working hours to the work ahead or adapt the work to the equipment and operators.

If there is *too much work* ahead, he can secure information from the Load Chart as to:

1. Deliveries to be quoted on future orders.
2. What kinds of orders must be refused.
3. Where congestion is likely to occur so that those processes can be studied and shortened.
4. What additional equipment to buy.

5. How many men to employ and the kind of work they will have
    to do.

6. What hours need to be lengthened.

If there is *not enough work* ahead, the manager can learn from the chart:

1. What kinds of orders are needed to keep the men or equipment
    busy (this information is the basis of sales or advertising cam-
    paigns, of reduction in prices, etc.).

2. What men to assign to other work.

3. What equipment can be disposed of.

4. What hours to shorten.

In order to furnish this information a Load Chart must be accurate

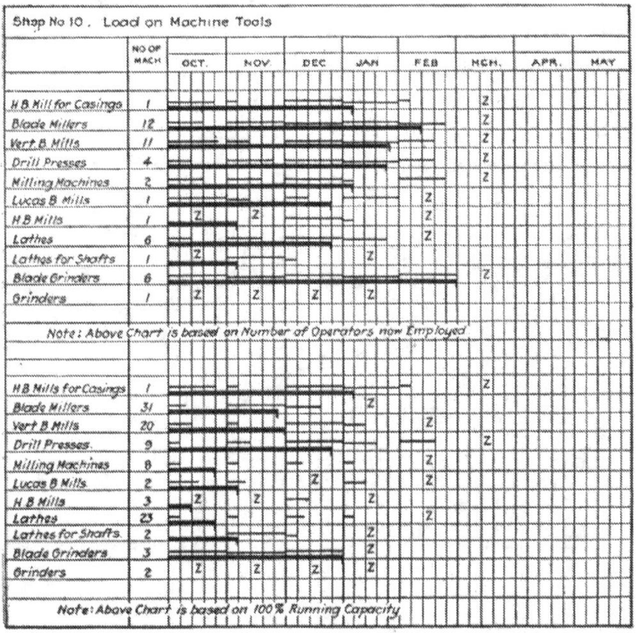

Fig. 10—A Gantt Load Chart Used in a Machine Shop

and up to date, but this is not difficult if it is based on Layout Charts
such as were described above.

The Gantt Load Chart shows very clearly whether or not the machine
tools in a plant are going to be kept busy in the near future, indicating which ones are overloaded and which have little work ahead. In
Fig. 10 the machine tools in shop No. 10 are listed in groups and the
lines show what part of the time they will be kept running to turn out
the orders in hand. This chart was drawn in a period of dull business and the information was, therefore, presented in two ways, the
first half of the chart listing only those machines to which operators
were assigned at that time and the second half listing all the machines

in the shop. The latter half, therefore, shows well in advance what machine tools will be idle unless more orders are secured, while the first half goes further and, in addition to telling what machines will be idle, shows what operators will have to be kept in the shop in idleness or be laid off if no more work is provided.

In this plant it was not necessary for the superintendent to spend hours in conversation with his foremen in an attempt to learn just how much work they had ahead of them, nor did he have to read long re-

Ordered Total 50,000	Size 7 9,915		Size 8 14,000		Size 9 4,335		Size 10 11 370		Size 11 7,775		Size 12 3,125	
Jos. Halsted Shoe Co. Date	Shipped on date	Shipped to and including date	Shipped on date	Shipped to and including date	Shipped on date	Shipped to and including date	Shipped on date	Shipped to and including date	Shipped on date	Shipped to and including date	Shipped on date	Shipped to and including date
May 18	1152	1152			370 950	1200						
May 21							810 954	1964				
May 22							912 948	3624				

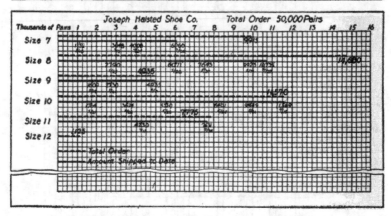

Fig. 11—Chart Showing Shipments vs. Orders

ports; on the Load Chart he had accurate information in condensed form.

Graphic Chart for Showing Progress in the Shipment of an Order.

An article by L. I. T. in *Factory*, November, 1919, illustrates a graphic chart by which the shipment of an order is pictorially recorded.

If you are shipping a portion at a time of some large order which requires several weeks or months to fill, probably you want to know occasionally how you are getting on. Perhaps, to give you the information, your clerk has to do some pencil work to find out.

This was the case in an eastern company manufacturing box toes, until the diagram and form shown in Fig. 11 were worked out.

In the case illustrated the Joseph Halsted Shoe Company has ordered

50,000 pair. Of these, 9,915 are to be of size 7; 14,680, size 8, and so on.

The form shown is typewritten and put in a loose-leaf book. Then the red lines (shown here as dotted) are drawn on a large sheet of cross-section paper which is mounted on a piece of wall board in the manager's office.

As shipments are made they are entered on the form in the left-hand column under the various sizes, and the cumulative figure to date put down in the right-hand column. This is well illustrated under size 10. Where two entries occur on the same date, as under sizes 9 and 10, this indicates two separate shipments on that day.

At the same time, the solid black line is extended on the chart in proportion to the size of the shipment and a notation made at the end of the line, showing the date and the exact number shipped to date. In this chart, which represents the standing of the order on the evening

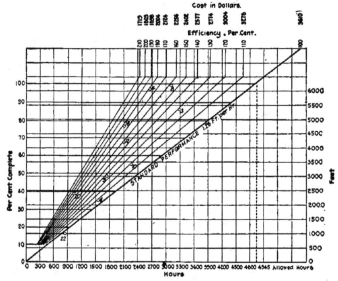

Fig. 12—Progress Chart—Electrical Conduit Installation

of July 27, shipments are completed on sizes 9, 10, and 11. On size 12, however, none is shipped. This setting down of the number makes it easy to add future shipments and serves as a check on the correctness of the chart.

This plan enables the manager to see at a glance the condition of all orders without separate calculation from the shipping records every time the question comes up.

A Chart Used as an Incentive to Production was given in an article by P. G. Salmon, Jr., in *Industrial Management*, February, 1920. The following notes are from Mr. Salmon's article.

A successful profit-sharing system must have a greater appeal than the monetary incentive offered to the workmen for extra productive effort. Individual performance must be shown in such a manner that the highest appeal is made to personal pride and the keenest competition encouraged.

To accomplish this a ratio chart has been designed to show the individual or gang performance, as the case may be. Such a chart should be kept where all the men may view it and compare their performance with others, which naturally will excite the keenest competition and therefore speed up production.

The chart, Fig. 12, is called an "active ratio chart of production" in that it records the progress of the work as it is being done and represents graphically the installation of electrical conduit on ships of exactly the same type, on a profit-sharing plan calling for the complete installation of 6,250 feet of conduit in 4,945 hours, subject to an agreement that any saving in hours is to be shared equally by the men and the company.

On the right side of the sheet the work or job is layed off in feet. On the lower margin the time allowed for the job is laid off in hours, so that a diagonal drawn from zero hours and feet to the point of intersection of the total job and total time allowed represents the set standard rate per hour of production, or, in other words, an efficiency of 100 per cent. The left margin is graduated to show per cent. of completion or progress and is laid off in convenient percentages.

The standard performance diagonal represents a production efficiency of 100 per cent. and the graduations on the upper magazine corresponding increases in efficiency or rates per hour and since it is agreed that savings are to be equally divided all production recorded over 100 per cent. efficient shows that the workmen increase their hourly rate 5 per cent. at 110 per cent. efficient, 10 per cent. at 120 per cent. efficient, and so on.

The production is reported daily for each hull and recorded on the chart using a small tab bearing a hull number and fastened to the chart with a push pin. The position on the chart of any hull is determined by the accumulative hours consumed, and feet of conduit installed since starting the work and the efficiency recorded at any position of a hull on the chart shows the men what increase they are earning over their hourly rate.

The direct cost of the total installation is shown at the several percentages of efficiency which furnishes the executive force a fair indication as to the final cost of an installation in the course of completion and, an excellent indication as to the relative merits of the foreman in charge of the work.

The principle involved in the ratio chart can be applied to any production work operated on a contract or agreement basis, where an allowance in hours or dollars is given for a unit of production. In actual use it has proven to be a great factor in keeping interest in the work at a high pitch and has given impetus to production.

The simplicity of the chart makes its use adaptable to almost any kind of factory output.

Scheduling a Mail Order Campaign. Fig. 13 shows a graphic chart for keeping a schedule of the sending out of form letters, etc. It was planned to send 10,000 letters during February. There were 22 working days in this month, so the average number of letters per day would be 455. A line was therefore drawn from 455 on the 1st day of the month to 10,000 for the 28th of the month. This straight line represents the rate at which the letters were to go out so that 10,000 would be sent by the end of the month. On the first day, 600 were actually sent; on the second none; on the third, 800; on the fourth, 400. The fifth was a Sunday and no letters were sent until the seventh. It will be seen

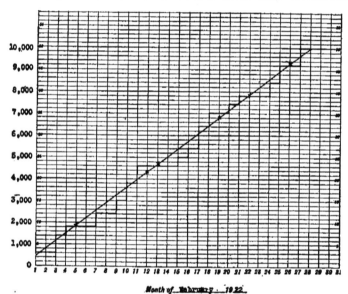

Fig. 13—Chart Schedule of Letters Sent Out

that on the sixth day of the month the schedule was 425 letters behind. On the seventh 600 letters were sent, but the schedule was still 180 behind.

The number of letters is plotted for each day for the whole period and the chart will show at a glance just how the actual performance is keeping up with the schedule. Sundays and holidays are indicated by crosses on the schedule line. Affecting conditions may be noted on the chart.

BIBLIOGRAPHY

BOOKS AND PAMPHLETS

The following is a list of books and pamphlets pertaining to the methods of collecting and presenting business statistics which have come to the authors' attention. All of them contain descriptions and illustrations, to a greater or less degree, on how to make and use graphic charts in business.

The publishers of this book have said that they will be glad to procure copies for their readers, at current prices.

Alexander, Carter	School Statistics and Publicity.
American Society of Mechanical Engineers: Joint Committee on Standards for Graphic Presentation	Preliminary Report.
Bailey, W. B. and Cummings, J.	Statistics
Berndt, Irving A.	Costs, Their Compilation and Use in Management
Bowley, A. L.	Elementary Manual of Statistics.
Brinton, W. C.	Graphic Methods for Presenting Facts
Cobey, J. W.	Traffic Field
Copeland, M. T.	Business Statistics
Dartnell Corporation	Graphic Charts for Picturing Sales Facts
Dartnell Corporation	Sales Managers' Desk Book
Duncan, C. S.	Commercial Research
Edgerton, E. I.	Business Mathematics
Ennis, W. D.	Cost Accounting
Eggleston, Dewitt C.	Works Management
Farnham, D. T.	Executive Statistical Control
Ficker, N. T.	Shop Expense Analysis and Control
Fisher, Irving	The "Ratio" Chart for Plotting Statistics
Frederick, J. G.	Business Research and Statistics
Frederick, J. G.	Modern Salesmanagement
Frost, P.	An Elementary Treatise on Curve Tracing
Gibson, G. A.	Elementary Treatise on Graphs
Gilman, Stephen	Graphic Charts for the Business Man
Gowin, E. B.	Developing Executive Ability.
Greenlinger, L.	Financial and Business Statements
Harvard University Committee on Economic Research	Indices of General Business Conditions by Warren M. Persons.—An Index of the Physical Volume of Production by Edward E. Day.—Review of Economic Statistics for 1919, 1920, 1921
Haskell, A. C.	How to Make and Use Graphic Charts
Jordan, J. P. and G. L. Harris	Cost Accounting Principles and Practice
King, W. I.	Elements of Statistical Methods
Knoeppel, C. E.	Graphic Production Control
Knoeppel, C. E.	Installing Efficiency Methods
Leffingwell, W. H.	Making the Office Pay
Marshall, W. C.	Graphical Methods
Palmer, A. R.	The Use of Graphs in Commerce and Industry
Rogers, L.	Newspaper Building

Runge, C. D. T.	Graphical Methods
Saliers, E. A.	Accounts in Theory and Practice
Secrist, H.	Introduction to Statistical Methods
Secrist, H.	Statistics in Business
Shaw Co., A. W.	Costs and Statistics
Shaw Co., A. W.	Graphic and Statistical Sales Helps
Shaw Co., A. W.	Organizing for Increased Sales
Sheldrake, T. S.	Harmsworth Business Library, Vol. 4
Whipple, G. C.	Vital Statistics
White, P.	Market Analysis
Yule, G. U.	An Introduction to the Study of Statistics

PERIODICALS

The following is a list of the more recent articles, which have come to the attention of the authors, containing information on the use of graphic charts in business. The name of the author, the title of the article, the name of the magazine in which the article appeared and the date of publication are given.

It is rather difficult to classify such articles satisfactorily as many of them are not limited to the discussion of one particular use of charts, but explain and illustrate their merits in connection with several different phases of business. However, in order to aid the reader as much as possible in finding material on the subjects in which he is especially interested, an endeavor has been made to arrange the material in some logical order.

The publishers have expressed their willingness to furnish the names and addresses of concerns to whom application may be made for the magazines listed.

The following articles pertain to line charts, both on plain and ratio, on the value of statistics in business, and especially graphic statistics:

Anderson, W. D.	Practical Use of Graphic Charts.	Business Methods, December, 1919.
Austin, O. P.	Use of Statistical Publications of the Government in Working out Problems of Commercial Investigation.	Administration, April, 1922.
Baldwin, G. P.	Commercial Statistics and their Value to the Executive.	General Electric Review, August, 1920.
Beaumont, R. H.	We Watch Time More Closely than Money.	System, June, 1921.
Bivins, P. A.	Charting as an Aid in Stabilizing Profits.	Industrial Management, May, 1922.
Burnet, A. R.	Trade Associations and Business Statistics.	Administration, December, 1921.
Burnet, A. R.	Comparability of Executive Control Charts.	Management Engineering, November, 1921.
Burnet, A. R.	Managing an Industry by Graphic Charts.	Management Engineering, August, 1921.
Cokell, W. B.	Statistics in Business.	Administration, May, 1921.
Conway. J. B.	Accumulative Graphic Record.	American Machinist, February 6, 1919.
DeLeeuw, A. L.	Applying "Moving Average" Charts to Industry.	Management Engineering, December, 1921.
Douglas, A. W.	Dealing with Business Facts.	Administration, January, 1922.

DuBrul, E. F.   Charting the Conditions of Business. Machinery, February, 1922.

Dye, G. M.   How Borrowed Ideas Helped us Grow. System, December, 1919.

Estes, L. V.   Visualizing Facts for Control.   Industrial Management, January and February, 1920

Farncombe, A.   Charts Simplify the Office Work.   Business Methods, May, 1921.

Farnham, D. T.   Industrial Administration.   Administration, May, 1922.

Flether, H. J.   How to Make a Report for the Chief Industrial Management. Executive.   January, 1921.

Fowler, E. J.   Value of Statistics in Central Station Electrical Review and Operation.   Industrial Engineer. November, 22, 1919.

Gary, E. H.   The Better Way to Plan Ahead.   System, May, 1921.

Hamburger, A. L.   Burden Absorption for Administra-   Administration, Septem-tive Control.   ber, 1921.

Haskell, A. C.   Graphic Charts for Office and Fac- India Rubber World, tory.   December 1, 1921.

Iron Trade Review   Curves Reveal Significant Facts.   October, 23, 1919.

Langley, S. S.   Graphic Representation of Oil Well Engineering & Mining Decline.   Journal, August 7, 1920.

Leland, F. H.   The Parallel Line of Control in   Management Engineering, Business.   December, 1921.

Mathewson, P.   Budgeting Business.   Industrial Management, June, 1921.

McKinsey, J. O.   Budgetary Control and Administra-   Administration, Janu-tion.   ary, 1921.

Mitchell, W. C.   How You Can Use The Business   System, December, 1921. Cycle.

Moody, R. E.   An Easy Way to Get Facts Straight. System, February, 1921.

Morrison, J. E.   Management in the Brass and Copper Management Engineer-Industry.   ing, February, 1922.

Parsons, F. W.   Everybody's Business: Graphic Meth- Saturday Evening Post, ods in Business.   May 27, 1922.

Phare, G. A.   Use of Graphs in Business Reports. Business Methods, March, 1920.

Phare, G. A.   Use of Graphs in the Factory.   Business Methods, June, 1920.

Rae, J.   Elements of Statistics.   Accountant, November, 27, 1920.

Richardson, G. H.   Business Statistics.   Accountant, March 1, 1919.

Rogers, J.   My Little Black Book.   Administration, Febru-ary, 1921.

Rorty, M. C.   Making Statistics Talk.   Industrial Management, December, 1920. January and February, 1921.

Scholfield, E. E.   Why We Don't Lose Letters.   System, June, 1919.

Schou, T.   Computing Power Factor Problems by Electrical Review and Graphic Methods.   Industrial Engineer, May 28, 1921.

Secrist, H.   Statistical Standards in Business Re- American Statistical As-search.   sociation Quarterly, March, 1920.

Stoddard, W. L.   The Use of Curves in Industry.   Factory, January, 1922.

Swarts, G. T., Jr. Drawing a Picture of Municipal Busi- American City, July, ness: Graphic Methods for City   1919. Officials.

Thomas P. E.   What Graphs Can Tell You About   System, January, 1922. Your Business.

U. S. Government   Publications of many of the depart-ments contain charts.

Von Huhn, R.	Capacity Bar Diagram.	Industrial Management, September, 1919.
Webb, A. D.	Introduction to the Elementary Terms and Methods of Statistics.	Incorporated Accountants Journal, August-October, 1920.
Woods, C. E.	Practical Organization of Industry.	Administration, May-June, 1921.

The following articles pertain to line charts, both on plain and ratio ruling:

Bivins, P. A.	Ratio Chart and its Application.	Industrial Management, July-October, 1921.
Burnet, A. R.	Standardizing the Z Chart.	Management Engineering, September, 1921.
Burnet, A. R.	Scale Selection for Z Chart.	Management Engineering, December, 1921.
Burnet, A. R.	Filing the Z Chart.	Management Engineering, January, 1922.
Caldwell, H.	Ratio Chart vs. Difference Chart.	Business Methods, March, 1920.

The two articles below are of interest in connection with the Probability Chart:

Cotton, H.	Method of Correlation as Applied to the Determination of Laws Followed by the Results of Experiment.	Electrician, April 2, 1920.
Hart, W. L.	Theory of Errors. Probability Curve.	American Waterworks Assn. Journal, November, 1920.

The articles below suggest uses for charts in accounting:

Bloor, W. F.	Value of Graphics in an Accounting System.	Journal of Accountancy, June, 1921.
Electrical World	Analysis of Consumption by Small Lighting Customers.	June 22, 1918.
Kurtz, E.	Replacement Insurance.	Administration, July, 1921.
Parsons, R. H.	Graphic Method of Keeping Continuous Power Plant Records.	Electrical Review and Industrial Engineer, February 21, 1920.

The use of charts in connection with advertising is described in the articles following:

Lambert, S. C.	Statistics-Using them Effectively in Advertising.	Printers Ink, August 11, 1921.
Richardson, A. H.	Graphic Methods in Technical Advertising.	Industrial Management, April, 1921.

Graphic Charts for the presentation and analysis of costs are described in the following articles:

Akerman, E. L. and Merrit, L. F.	Graphic Analysis of Manufacturing Costs.	American Industries, July, 1919.
Balsam, L.	Do Your Letters Cost Too Much?	System, November, 1919.
Barness, F. E.	Curves of Building Cost Increase Used in Valuation Work.	Engineering News-Record, October 7, 1920.
Bonner, H. R.	Graphic Method for Presenting Comparative Cost Analysis.	American Statistical Assn.. Quarterly, September, 1920.
Carr, J. B.	How Graphics Help in Estimating Costs.	Industrial Management, January, 1921.
Cole, D. S.	Estimating "Cost Promises" in Small Plant.	Industrial Management, March, 1922.
Conner, W. N.	Our Men Know All About Our Costs.	System, May, 1920.
Goldthwait, C. F.	Graphic Method for Textile Calculations.	Textile World, December 31, 1921.
Herlihy, F. J.	Charts Tell Cost of Concrete Paving.	Engineering News-Record, February 5, 1920.
Hood, K. K.	Curves for Ore Valuation.	Mining & Scientific Press, August 21, 1920.
Myers, D. M.	Cost Cutting for Industrial Power Plants.	Industrial Management, May, 1922.
Patch, C. E	Graphic Building Costs.	Engineering & Contracting, July 28, 1920.
Tomlinson, M. C. W.	The Influence of Weather on Coal Consumption.	Management Engineering, December, 1921.
Vincent, G. I.	Graphic Method of Cost Analysis.	American Gas Association Monthly, August, 1919.

The articles following illustrate the value of graphic charts for visualizing general business and financial facts:

Ayres, L. P.	Price Changes and Business Prospects.	Administration, August, 1921.
Burnet, A. R.	A Chart to Take to the Bank.	Administration, November, 1921.
Crabtree, J. A.	How Much Capital Does a Business Really Need?	System, November, 1911.
Factory Magazine	International Industrial Digest in each number. Charts.	
Hutchins, F. L.	Graphic Study of Federal and Private Railroad Management.	Annalist, April 25, 1921.
Management Engineering Magazine.	Each issue has charts accompanying data on The Labor Summary; Statistics and News from Industry, and Industrial Index Numbers.	
Morgan, J. D.	Efficient Operation of Central Power Stations.	Power, October 7, 1919.
System Magazine.	Recent issues have chart accompanying article on Trend of Business.	

Suggestions for the use of charts for inventory control are given in the articles below:

Holmes, B. E.	The Ratio Chart Applied to Inventory Control.	Industrial Management, April, 1922.
Woods, C. E.	Elements of Inventory Control.	Administration, August, 1921.

The value of graphic methods in organization and management is related in the following articles:

Bigelow, C. M.	The Organization of Knitting Mills.	Management Engineering, January, 1922.
Boffey, L. F.	Basic Principles of Stores Organization.	Administration, January, 1921.
Fisher, B.	Charting Authority and Responsibility to Show Complex Management Relationships.	Management Engineering, May, 1922.
Fitting, R. U.	Organization for Construction Work.	Administration, August, 1921.
Howard, N.	Industrial Organization Applied to Ship Yard Management.	Industrial Management, July, 1920.
Merry, H. M.	Graphic Metallurgical Control.	American Institute of Mining & Metallurgical Engineers Bulletin, September, 1919.
Munn, G. G.	Charting the Organization.	Administration, March, 1921.
Van Deventer, J. H.	Planning Department Systems: How to Visualize Methods by Mapping the Routine.	Industrial Management, November, 1920.

Graphic charts in the personnel department for presenting payment plans and for visualizing labor turnover, the results of training employees, causes of accidents, etc., are described in the following articles:

Anderson, W. D.	Modern Forms of Wage Systems.	Business Methods, February, 1920.
Benge, E. J.	A Department of Personnel Research.	Industrial Management, May, 1920.
Benge, E. J.	How to Analyze the Working Force.	Management Engineering, February, 1922.
Bober, W. C.	Graphic Planning of Payroll Procedure.	Industrial Management, November, 1920.
Brissenden, P. F. and Frankel, E.	Causes of Labor Turnover.	Administration, November, 1921.
Bundesman, C. S.	Value of the Personal Record.	Industrial Management, November, 1919.
Crowther, S.	How to Handle Wage Cuts.	System, May, 1921.
Faltin P. and Blog, L.	Wage Payment Administration and its Relation to Production Control.	Industrial Management, August, 1920.
Harrison, G. C.	The High Cost of Idleness.	Management Engineering, August, 1921.
Heilman, R. E.	Where Profit Sharing Pays Best.	System, February, 1920.
Johnson, J. F.	Suggestions Regarding Factory Training.	Industrial Management, February, 1920.
Johnson, J. F.	Training as a Factor in Reducing Waste.	Management Engineering, August, 1921.
Keeley, H. L.	Placing Responsibility for Accidents.	Management Engineering, March, 1922.
Kenagy, H. G.	The Prevention of Labor Turnover.	Administration, October, 1921.
Lichtner, W. O	How To Work Up & Use Time Studies.	Industrial Management, August, 1920.
Lucey, E. A.	The Cost of Mismanagement.	Management Engineering, January, 1922.

Martindale, J. J.	A Bonus Plan that Rewards Economy.	Sales Management, May, 1922.
May, H. B.	The Part Inspection Plays in Good Management.	Factory, January 15, 1921.
Morgan, E. S.	Practice Personnel Management.	Industrial Management, May, 1920.
Osgood, M. W.	Bigger Wages, More Work and Fewer Employees.	System, August, 1919.
Otis, A. S.	Determination of the Optimum Number of Machines a Workman Should Operate.	Industrial Management, December, 1919.
Polakov, W. N.	The Measurement of Human Work.	Management Engineering, February, 1922.
Reeves, H. H.	An Analysis of Turnover.	Administration, October, 1921.
Sedgwick, J. R.	How We Cut Training Costs.	Factory, January 1, 1921.
Swartz, G. O.	The Power of Pride.	Industrial Management, March, 1921.
Sylvester, L. A.	Measuring the Human and Mechanical Elements in Manual Work.	Industrial Management, September, 1921.
Wells, W. S.	Visualizing Employment Records	Industrial Management, July, 1920.
Wolf, R. B.	Making Men Like Their Jobs.	System, February, 1919.

The following article illustrates admirably the important part charts may play in planning ahead:

Scoville, J. W.	How We Plan Four Months Ahead.	Factory, July, 1921.

Some of the great many uses of charts in the sales department are described in the articles below:

Burnet, A. R.	A Simple Method of Charting Sales.	Administration, September, 1921.
Clark, W.	Is It Profitable to Sell At a Loss?	Factory, March, 1922.
Hawes, E. M.	Better Results from Maps.	System, July, 1920.
Howell, A. R.	Facts from Which to Chart the Pulse of Selling.	Printers Ink, January, 22, 1920.
Hyde, J. R.	My Method of Setting Quotas.	System, March, 1920.
McPherson, C.	More Life for the Sales Manual.	Sales Management, December, 1921.
Peck, E. C.	How Our Sales Records Tell Us What to Buy.	Factory, June 1, 1921.
Polakov, W. N.	Distributing Overhead to Allow Lower Sales Prices.	Factory, April, 1922.
Selmor	Maps and Tacks as an Aid to Sales.	Business Methods, April, 1920.
Tuttle, W. F.	Cutting Territories to Sell More.	System, December, 1919.

Charts for production control—scheduling and progress—are described and illustrated in the articles following:

Anderson, W. D	Checking Production through the Plant.	Business Methods, December, 1919.
Basset, W. R.	Modern Production Methods: Graphic Methods of Control.	American Machinist, September 15, 1921.
Brown, R. T.	Graph Records Progress of Road-Survey Parties.	Engineering News-Record, February 12, 1920.
Clark, W.	The Gantt Chart	Management Engineering, August-November, 1921.

Daniels, F. R.   Production Control by Graphics.   Machinery, October, 1921.

Electrical World   Compact Central Station Chart to Record Progress.   January 12, 1918.

Fontaine, J. E.   Graphical Records for Progress in Highway Surveys.   Engineering News-Record, July 8, 1920.

Hyde, W. R.   Using Graphic Charts to Increase Pressroom Production.   Inland Printer, March, 1921.

Penrose, C.   Planning and Progress on a Big Construction Job.   Engineering News-Record, March 18 and 25, 1920.

Rahn, R. J.   Graphic Production Control for the Small Shop.   Industrial Management, June, 1921.

Salom, P. G., Jr.   Ratio Chart as Incentive to Production.   Industrial Management, February, 1920.

Schmidt, L. W.   Enlisting Labor in Production.   American Machinist, February, 1919.

Spidy, E. T.   Graphic Production Control in Railway Shops.   Railway Mechanical Engineer, April, 1920.

Trabold, F. W.   How a Manager Uses Gantt Charts.   Management Engineering, January, 1922.

INDEX

HOW TO MAKE AND USE
GRAPHIC CHARTS

BY
ALLAN C. HASKELL

With an Introduction by
RICHARD T. DANA

This book covers a somewhat different field than Graphic Charts in Business.

While it is, in no sense. a complete theoretical treatise, it is rather more technical than the latter.

It was written from the viewpoint of the engineer rather than from that of the business man, and while it takes up the discussion of graphic charts in connection with many subjects of as much interest to one as to the other, the majority of the examples and illustrations are drawn from the engineering field.

It is recommended as an excellent follow-up for Graphic Charts in Business for, given an understanding of the principles, it is but a step further to master some of the more complicated charts, such as the logarithmic chart, the alignment chart, etc., and to adapt them to the solution of business problems as well as engineering ones.

Many of the subjects covered in this book, such as Cost Analysis; Operating Characteristics; Results of Tests and Experiments; Computation, Arithmetical and Geometrical; Designing and Estimating; etc., are not touched upon in Graphic Charts in Business.

Send for a copy of "How to Make and Use Graphic Charts" on 10 days' approval.

CODEX BOOK COMPANY, Inc.,
119 Broad Street, New York, N. Y.